John Vince

Imaginary Mathematics
for Computer Science

 Springer

John Vince
Bournemouth University
Poole, UK

ISBN 978-3-319-94636-8   ISBN 978-3-319-94637-5   (eBook)
https://doi.org/10.1007/978-3-319-94637-5

Library of Congress Control Number: 2018949636

This Springer imprint is published by the registered company Springer Nature Switzerland AG
The registered company address is: Gewerbestrasse 11, 6330 Cham, Switzerland

This book is dedicated to my wife He^{i}diπ.

Preface

I first came across the $\sqrt{-1}$ in the complex roots of quadratic equations. i, or its doppelgänger j, popped up in my electrical engineering studies, where it separates phase differences in voltages and currents. In mathematics, I learned about Euler's equation $e^{i\pi} + 1 = 0$ and how i creates totally new subjects, such as *complex function theory*. In industry, I came across *quaternions*, which are complex numbers in four dimensions. More recently, I have discovered *octonions* and *geometric algebra*. This journey of discovery has been long and arduous, but exciting.

During my youth, I questioned the meaning of i, but I no longer worry about such matters. However, I wish that I had discovered all that I now understand about imaginary mathematics from one source, which is the reason behind this book. I remember trying to understand an internal document on quaternions during my time in flight simulation. I felt that the author had written the document to deliberately hide the contents from me. I learned nothing from this communication, apart from a determination to understand the subject.

Although I am far from being an expert in mathematics, I would like to pass on what I have discovered about complex numbers in the following chapters. I suppose I had to include an obligatory introductory chapter tracing the history of i's rise to fame. Chapter 2 on *Complex Numbers* places them in a numerical context and describes topics such as the complex plane, complex exponentials, logarithms, hyperbolic functions and simple derivatives. I have included many illustrations and worked examples to reinforce the mathematical ideas.

Chapter 3 is on *Matrix Algebra* and describes topics such as complex eigenvalues and eigenvectors, representing complex numbers as matrices, complex matrix algebra, and the complex inner and outer products. I also include many worked examples.

Quaternions are the subject of Chap. 4, and take my word, that if you understand complex numbers, then quaternions are just as easy. The chapter starts with Hamilton's struggle to develop a 3D form of complex numbers, describes the various forms and associated algebra, and concludes with some worked examples.

Octonions are new to me, and Chap. 5 reveals what I have discovered from my research. The Cayley–Dickson construction shows that an octonion can be regarded as an ordered pair of quaternions; a quaternion is an ordered pair of complex numbers, and a complex number is an ordered pair of reals. Even if you never use them in your work, at least you know where they belong in imaginary mathematics.

Chapter 6 describes geometric algebra, which was not developed to exploit the imaginary unit, but turns out to possess imaginary qualities. I have previously written about the subject and believe that it will play an important role in future descriptions of science and physics. I describe the various products associated with different geometric elements and their relationship to quaternions.

The rest of the book deals with applications of the above algebras. Chapter 7 shows how complex numbers simplify the representation of compound angles, and Chap. 8 describes how complex exponential notation simplifies the combination of waves. This chapter shows the importance of complex numbers in dealing with wave phenomena, be they simple water waves or waves in quantum fields.

Chapter 9 covers *Circuit Analysis Using Complex Numbers*. The objective is not to turn you into an electrical engineer, but to reinforce the role of complex notation in representing out-of-phase electrical waves.

Chapter 10 is on *Geometry Using Geometric Algebra* and may inspire you to write software using GA's constructs. Still on a geometric theme, Chap. 11 shows how quaternions are used to rotate vectors about an arbitrary 3D axis.

I have always been fascinated by prime numbers, especially the Riemann hypothesis. Entire books have been written on the subject, and in Chap. 12, I have attempted to condense the explanation to half-a-dozen pages. Chapter 13 describes the simple algorithm behind the Mandelbrot set, using some beautiful images provided by Dr. Wolfgang Beyer and Dr. Dominic Ford.

The last chapter concludes the book and reminds the reader how complex numbers have found their way into quantum physics, by including references to Pauli matrices, Dirac matrices, the Dirac equation, and the Schrödinger equation.

I have really enjoyed writing and researching this book. During this time, I have discovered some extremely well-written books and articles on the Internet. As always, Wikipedia is an amazing resource, and long may it continue as an independent agency. I thank Dr. Tony Crilly for reading the final manuscript and making some important suggestions. Naturally, if I have included any mistakes, they are of my own doing!

As always, I thank Beverley Ford, Editorial Director—Computer Science, and Helen Desmond, Editor—Computer Science for Springer-Verlag, for the support and guidance they have provided throughout the book's development.

Finally, enjoy this fascinating subject.

Breinton, Herefordshire, UK   John Vince
August 2018

Contents

Chapter 1
Introduction

1.1 Why i is Necessary

Early civilisations only required to count using positive integers, making arithmetic relatively easy. However, with the advent of negative numbers new rules had to be found. The Indian mathematician and astronomer Brahmagupta (598-c.–670), showed how positive and negative numbers interacted with one-another, and proposed the rules in Table 1.1. Table 1.2 shows the rules for multiplying and dividing positive and negative numbers.

Children and adults are often surprised that the product or division of two negative numbers results in a positive number, but a little algebra proves why this must be.

Consider the expansion of $(a + b)^2$:

$$(a + b)^2 = (a + b)(a + b)$$
$$= a^2 + 2ab + b^2.$$

When $a = 5$ and $b = 2$, then

$$(5 + 2)^2 = (5 + 2)(5 + 2)$$
$$= 25 + 2 \times 5 \times 2 + 4$$
$$= 49$$

which is correct.

Now consider the expansion of $(a - b)^2$:

$$(a - b)^2 = (a - b)(a - b)$$
$$= a^2 - 2ab + b^2.$$

When $a = 5$ and $b = 2$, then

© Springer International Publishing AG, part of Springer Nature 2018
J. Vince, *Imaginary Mathematics for Computer Science*,
https://doi.org/10.1007/978-3-319-94637-5_1

Table 1.1 Rules for adding and subtracting positive and negative numbers

+	b	$-b$
a	$a+b$	$a-b$
$-a$	$b-a$	$-(a+b)$

$-$	b	$-b$
a	$a-b$	$a+b$
$-a$	$-(a+b)$	$b-a$

Table 1.2 Rules for multiplying and dividing positive and negative numbers

$\times$	b	$-b$
a	ab	$-ab$
$-a$	$-ab$	ab

$/$	b	$-b$
a	a/b	$-a/b$
$-a$	$-a/b$	a/b

$$(5-2)^2 = (5-2)(5-2)$$
$$= 25 - 2 \times 5 \times 2 + 4$$
$$= 9$$

which is also correct, and assumes: $-2 \times -2 = +4$. The rule we learn as children: "two negatives make a positive" ensures that algebra is consistent. Consequently, for any x, $(\pm x)^2 \geq 0$, which implies that $\sqrt{-x}$ cannot have a numerical solution. However, this does not prevent us from inventing a symbol that breaks this rule: $i^2 = -1$; it simply means that i is not an ordinary number.

1.2 The Language of Mathematics

Mathematics is a language for describing problems involving numbers and unknown quantities. For example, I can express a numerical problem as follows: What number, when squared and increased by 13 equals 49? Using algebra this becomes

$$x^2 + 13 = 49$$

which when manipulated reveals

$$x^2 = 36$$

and provides the result:

$$x = \pm 6.$$

Fig. 1.1 Graph of $y = x^2 - 3x + 2$

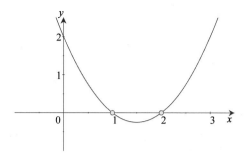

I am certain that most readers solved this problem without algebra, but algebra becomes very handy as the question becomes more elaborate, such as: What number, when squared and reduced by three times the original number equals -2? Using algebra this is

$$x^2 - 3x = -2$$

and is solved as follows

$$x^2 - 3x = -2$$
$$x^2 - 3x + 2 = 0$$
$$(x - 1)(x - 2) = 0$$

and reveals:

$$x = 1, \quad x = 2.$$

Figure 1.1 illustrates the two roots.

Now let's modify the question slightly: What number, when squared and reduced by three times the original number equals -3? Algebraically, this is

$$x^2 - 3x = -3 \tag{1.1}$$

and is solved as follows

$$x^2 - 3x = -3$$
$$x^2 - 3x + 3 = 0$$

and makes us think about a possible solution.

Fortunately, we know that an equation of the form

$$ax^2 + bx + c = 0$$

has roots

$$x = \frac{-b \pm \sqrt{b^2 - 4ac}}{2a}$$

which when applied to (1.1) reveals

$$x = \frac{3 \pm \sqrt{9 - 12}}{2}$$
$$= \frac{3 \pm \sqrt{-3}}{2}. \tag{1.2}$$

We could stop at this point as there is no real solution to $\sqrt{-3}$, which means that there is no real solution to the original question. However, simply by introducing the subterfuge $i^2 = -1$ into (1.2) we have

$$x = \frac{3 \pm \sqrt{3i^2}}{2}$$
$$= \frac{3 \pm \sqrt{3}\, i}{2}$$
$$= 1.5 \pm \sqrt{0.75}\, i$$

which expresses the solution in terms of a new object, i. There is still no real solution to (1.1), but we have found what are called the *complex roots* to the original problem, which may not seem very useful.

Figure 1.2 shows why (1.1) has no real roots: the parabola never touches or intersects the x-axis.

Just to make sure, let's substitute these complex roots back into (1.1).

Substituting $x = 1.5 + \sqrt{0.75}\, i$ into (1.1)

Fig. 1.2 Graph of $y = x^2 - 3x + 3$

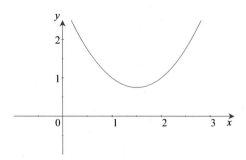

$$x^2 - 3x = \left(1.5 + \sqrt{0.75}\,i\right)^2 - 3\left(1.5 + \sqrt{0.75}\,i\right)$$
$$= 2.25 + 3\sqrt{0.75}\,i + 0.75\,i^2 - 4.5 - 3\sqrt{0.75}\,i$$
$$= -2.25 - 0.75$$
$$= -3$$

which is correct. Similarly, substituting $x = 1.5 - \sqrt{0.75}\,i$ into (1.1)

$$x^2 - 3x = \left(1.5 - \sqrt{0.75}\,i\right)^2 - 3\left(1.5 - \sqrt{0.75}\,i\right)$$
$$= 2.25 - 3\sqrt{0.75}\,i + 0.75\,i^2 - 4.5 + 3\sqrt{0.75}\,i$$
$$= -2.25 - 0.75$$
$$= -3$$

which is also correct.

The addition of i to the language of mathematics illustrates how mathematics evolves over time. i is now recognised as a fundamental feature of mathematics, and when we ask "but what is i?" we can only reply, it is not a number, but i is a symbol with the property that $i^2 = -1$. It is often called the *imaginary unit*. Soon we will see that i behaves like a spatial operator, and often reveals amazing hidden patterns and symmetries.

1.3 A Brief History of i

The Bologna mathematician Scipione del Ferro (1465–1526), was particularly interested in solving cubic equations. For example, Fig. 1.3 shows the graph of the cubic function $y = x^3 - 6x^2 + 11x - 6$ which has three distinct, real roots $x = 1, 2, 3$. Figure 1.4, on the other hand, shows the graph of $y = x^3 - 5x^2 + 8x - 4$, where the root $x = 2$ occurs twice.

Fig. 1.3 Graph of
$y = x^3 - 6x^2 + 11x - 6$

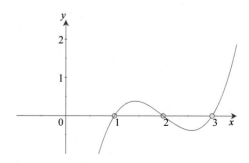

Fig. 1.4 Graph of
$y = x^3 - 5x^2 + 8x - 4$

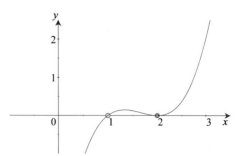

Fig. 1.5 Graph of
$y = x^3 - 5x^2 + 9x - 5$

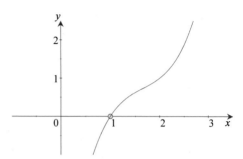

Figure 1.5 shows the graph of $y = x^3 - 5x^2 + 9x - 5$ with a single real root $x = 1$, and two complex roots. Although del Ferro could not call upon complex numbers, he was aware that cubics could have one, two and three real roots.

The Italian polymath Gerolamo Cardano (1501–1576), continued del Ferros' interest in cubics and acknowledged the existence of terms such as $\sqrt{-23}$, but considered them "sophistic" [1], plausible but fallacious.

The Italian mathematician Rafael Bombelli (1526–1572), developed Cardano's techniques for solving cubic equations, and although he showed that expressions involving the square-root of negative terms obeyed the laws of algebra, he was unable to find a physical meaning for $\sqrt{-1}$.

The French philosopher, mathematician and scientist René Descartes (1596–1650), published *La Géométrie* in 1637, and in the context of the roots for the cubic

$$y^3 - 3a^2y + \frac{3a^3c^3}{b^3}\sqrt{3} = 0$$

he commented:

Neither the true nor the false roots are always real; sometimes they are imaginary; (Mais quelquefois seulment imaginaires) .... Thus, while we may conceive of the equation $x^3 - 6x^2 + 13x - 10 = 0$ as having three roots, yet there is only one real root, 2, while the other two, however we may increase, diminish, or multiply them in accordance with the rules just laid down, remain always imaginary [2].

Fig. 1.6 Graph of
$y = x^3 - 6x^2 + 13x - 10$

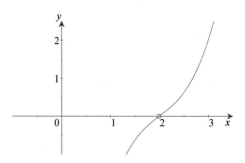

Figure 1.6 shows the graph of the equation referred to by Descartes. Unfortunately, the word *imaginary* is today associated with ideas or thoughts created by our imagination, such as *ghosts* or *fairies*, whereas the $\sqrt{-1}$ is far from imaginary. It has been suggested that in an historical context, imaginary was a derogatory term [3].

The English mathematician John Wallis (1616–1703), was one of the leading mathematicians of the seventeenth century, and in his treatise *Algebra* he anticipated the concept of complex numbers, i.e. $a + bi$. Wallis is also credited with inventing the idea of the number line, with positive numbers to the right and negative numbers to the left.

The French mathematician Abraham de Moivre (1667–1754), is particularly remembered for a formula published in 1722 – De Moivre's formula:

$$(\cos\theta + i\sin\theta)^n = \cos(n\theta) + i\sin(n\theta).$$

In 1749 the Swiss mathematician, physicist, astronomer, logician and engineer Leonhard Euler (1707–1783), used his own formula

$$e^{i\theta} = \cos\theta + i\sin\theta$$

to prove De Moivre's formula for any real *n*. Euler wrote in his *Elements of Algebra* of 1770:

> All such expressions, as $\sqrt{-1}$, $\sqrt{-2}$, $\sqrt{-3}$, $\sqrt{-4}$, &c. are consequently impossible, or imaginary numbers, since they represent roots of negative quantities; and of such numbers we may truly assert that they are neither nothing, nor greater than nothing, nor less than nothing; which necessarily constitutes them imaginary, or impossible [4].

Even though Euler regarded terms involving $\sqrt{-1}$ as *imaginary*, he did show that $e^{ix} = \cos x + i\sin x$, and when $x = \pi$ we get $e^{i\pi} + 1 = 0$, which has become such an icon in mathematics.

As the words *imaginary* and *impossible* had started to be associated with $\sqrt{-1}$, Euler took the first letter of imaginary and adopted the notation $i = \sqrt{-1}$ [5]. And in anticipation of Jean-Robert Argand's use of the complex plane, Euler visualised complex numbers as points with rectangular coordinates.

The Norwegian-Danish mathematician Caspar Wessel (1745–1818), used complex numbers to represent vectorial quantities, and published a paper in 1799 describing complex numbers and the complex plane: *Om Directionens analytiske Betegning, et Forsog, anvendt fornemmelig til plane og sphriske Polygoners Oplosning*, but because it was in Danish it remained undiscovered until 1897 [6]. Today, Wessel is universally recognised as the inventor of the complex plane for describing complex numbers. One of the world's foremost theoretical physicists Roger Penrose, refers to the complex plane in his book *Fashion, Faith and Fantasy in the New Physics of the Universe* as the *Wessel Plane* [7].

The Swiss amateur mathematician Jean-Robert Argand (1768–1822), was unaware of Wessel's paper when he published his own paper in 1806 on a graphical basis for complex numbers: *Essai sur une manière de représenter des quantités imaginaires dans les constructions géométriques* [8]. Even though Wessel published before Argand, Argand's name is today associated with the complex plane – Argand plane. Argand also noted that when a complex number of the form $a + bi$ is multiplied by i, it is rotated 90° about the origin to a new point $-b + ai$.

The German mathematician Carl Friedrich Gauss (1777–1855), formalised and disseminated the geometric interpretation of i, and in his 1831 publication *Theoria Residuorum Biquadraticorum*, introduced the term *complex number*. In the second memoir (*Werke* 2) he wrote:

> If this subject has hitherto been considered from the wrong viewpoint and thus enveloped in mystery and surrounded by darkness, it is largely an unsuitable terminology which should be blamed. Had $+1$, -1 and $\sqrt{-1}$, instead of being called positive, negative and imaginary (or worse still, impossible) unity, been given the names say, of direct, inverse and lateral unity, there would hardly have been any scope for such obscurity [9].

The French mathematician and physicist Baron Augustin-Louis Cauchy (1789–1857), is remembered for several contributions to mathematics and physics, but in the context of complex numbers he is recognised for developing *complex function theory*.

The brilliant Irish mathematician, physicist and astronomer Sir William Rowan Hamilton (1805–1865), is particularly known for inventing *quaternions*, which are a four-dimensional complex number, and take the form $a + ri + sj + tk$, where $i^2 = j^2 = k^2 = ijk = -1$. Hamilton published *On Quaternions: Or a New System of Imaginaries in Algebra* in 1844 [10].

Hamilton invented quaternions in October 1843, and by December of the same year, his friend, Irish mathematician John Thomas Graves (1806–1870), had invented *octaves*, an eight-dimensional complex number, which would eventually be called *octonions*. The British mathematician Arthur Cayley (1821–1895), had also been intrigued by Hamilton's quaternions, and independently invented octonions in 1845. Octonions eventually became known as *Cayley numbers* rather than *octaves*, simply because Graves did not publish his results until 1848 – three years after Cayley!

Hamilton also originated the algebra of ordered pairs of numbers, and showed that complex numbers could be regarded as a pair of numbers. In addition to complex numbers, quaternions occupy a central place in mathematical systems, and today there are four such composition algebras: real $\mathbb{R}$, complex $\mathbb{C}$, quaternion $\mathbb{H}$, and octonion $\mathbb{O}$ that obey an n-square identity used to compute their norms.

In 1898 the German mathematician Adolf Hurwitz (1859–1919), proved that the product of the sum of n squares by the sum of n squares is the sum of n squares only when n is equal to 1, 2, 4 and 8, which are represented by the reals, complex numbers, quaternions and octonions. This is known as *Hurwitz's Theorem*, or the *1, 2, 4, 8 Theorem*. No other system is possible, which shows how important quaternions are within the realm of mathematics. Consequently, Hamilton's search for a system of triples was futile, because there is no three-square identity.

1.4   Summary

From the above, one can see how long it has taken for the $\sqrt{-1}$ to be taken seriously. Descartes' use of the word "imaginary" was unfortunate, and Gauss was correct to comment that the subject was "enveloped in mystery and surrounded by darkness" due to this label. However, we are where we are, and there is no way we can rewrite history or change the nomenclature. i will remain an imaginary quantity, and $a +
bi$ will always be a complex number, even though they are neither imaginary, nor complex.

To help the reader understand the role of complex numbers within the world of mathematics and science, I will conclude this chapter with a quote by Roger Penrose:

> However, the terminology is misleading, for it suggests that there is some greater "reality" to these so-called real numbers than there is to the so-called imaginary numbers. The impression comes about, I suppose, because there is the feeling that distance measures and time measures are, in some sense "really" such real-number quantities. But we do not know this. We know that these real numbers are indeed very good for describing distances and times, but we do not know that this description holds good at absolutely *all* scales of distance or time. We have no actual understanding of the nature of a physical continuum at a scale of, say, one googolth of a metre or of a second, for example. The so-called real numbers are *mathematical* constructions, which are, nevertheless, immensely valuable for the formulation of the physical laws of classical physics.

> Yet, real numbers may also be regarded as "real" in the *Platonic* sense – the same Platonic sense as any other consistent mathematical structure – if we are to adopt the common stand-point among mathematicians whereby mathematical consistency is the sole criterion for such Platonic "existence". However, the so-called imaginary numbers form just as consistent a mathematical structure as do the so-called real numbers, so, in this Platonic sense, they are also just as "real". A separate (and, indeed, *open*) question is the extent to which either of these number systems precisely models the actual world [7].

References

1. Nahin PJ (1998) An imaginary tale: the story of $\sqrt{-1}$. Princeton University Press, Princeton
2. Descartes R (1637) La Géométrie, (There is an English translation by Michael Mahoney). Dover, New York (1979)
3. Martinez AA (2006) Negative math: how mathematical rules can be positively bent. Princeton University Press, Princeton
4. Euler L (1770) Elements of algebra, Translated from the French by Francis Horner, M. Bernoulli and M. de la Grange, 1822, books.google.co.uk
5. Dunham W (1999) Euler, the master of us all, The Dolciani mathematical expositions, Number 22, Mathematical association of America
6. Wessel C (1799) Om Directionens analytiske Betegning, et Forsog, anvendt fornemmelig til plane og sphriske Polygoners Oplosning [On the analytic representation of direction, an effort applied in particular to the determination of plane and spherical polygons]. Nye Samling af det Kongelige Danske Videnskabernes Selskabs Skrifter (in Danish). Copenhagen: R. Dan. Acad. Sci. Lett. 5: 469518
7. Penrose R (2016) Fashion faith and fantasy in the new physics of the universe. Princeton University Press, Princeton, p 56
8. Argand JR (1874) Essai sur une manière de représenter des quantités imaginaires dans les constructions géométriques, 2nd edn. Gauthier-Villars, Paris
9. Ebbinghaus H et al (2012) Numbers. Springer, Berlin, p 61
10. Hamilton W R (1844) On quaternions: or a new system of imaginaries in algebra. Phil. Mag. 3rd ser. 25

Chapter 2
Complex Numbers

2.1 Introduction

In this chapter I review the axioms associated with different number systems, and show how they also cover imaginary and complex numbers. The complex plane is described as a way of visualising complex numbers and various algebraic operations, and two functions for isolating the real and imaginary parts of a complex expression. The section on Complex Algebra examines topics such as the complex conjugate, powers of i, complex exponentials, logarithms of a complex number, and the hyperbolic functions. Finally, there are a dozen worked examples.

2.2 Laws of Algebra

Laws or *axioms* provide a formal basis for mathematics, and in the following descriptions a *binary operation* is an arithmetic operation such as $+$, $-$, $\times$, $/$ which operate on two operands.

2.2.1 Commutative Law

The *commutative law* in algebra states that when two numbers are associated with certain binary operations, the result is independent of the order of the numbers. The commutative law of addition is

$$a + b = b + a$$
$$\text{e.g. } 1 + 2 = 2 + 1 = 3.$$

The commutative law of multiplication is

© Springer International Publishing AG, part of Springer Nature 2018
J. Vince, *Imaginary Mathematics for Computer Science*,
https://doi.org/10.1007/978-3-319-94637-5_2

$$a \times b = b \times a$$
$$\text{e.g. } 1 \times 2 = 2 \times 1 = 2.$$

Note that subtraction is not generally commutative:

$$a - b \neq b - a$$
$$\text{e.g. } 1 - 2 \neq 2 - 1.$$

2.2.2   Associative Law

The *associative law* in algebra states that when three or more numbers are associated with certain binary operations, the result is independent of how each pair of numbers is grouped. The associative law of addition is

$$a + (b + c) = (a + b) + c$$
$$\text{e.g. } 1 + (2 + 3) = (1 + 2) + 3 = 6.$$

The associative law of multiplication is

$$a \times (b \times c) = (a \times b) \times c$$
$$\text{e.g. } 1 \times (2 \times 3) = (1 \times 2) \times 3 = 6.$$

However, note that subtraction is not generally associative:

$$a - (b - c) \neq (a - b) - c$$
$$\text{e.g. } 1 - (2 - 3) \neq (1 - 2) - 3.$$

It turns out that octonions are neither commutative nor associative.

2.2.3   Distributive Law

The *distributive law* in algebra describes an operation which when performed on a combination of numbers is the same as performing the operation on the individual numbers. The distributive law does not work in all cases of arithmetic. For example, multiplication over addition holds:

$$a(b + c) = ab + ac$$
$$\text{e.g. } 2(3 + 4) = 6 + 8 = 14$$

whereas addition over multiplication does not:

$$a + (b \times c) \neq (a + b) \times (a + c)$$
$$\text{e.g. } 3 + (4 \times 5) \neq (3 + 4) \times (3 + 5).$$

Although these laws are natural for numbers, they do not necessarily apply to all mathematical objects. For instance, vectors and matrices do not commute.

2.3 Types of Numbers

Here are the various types of numbers and their set names.

2.3.1 Natural Numbers

The *natural* numbers $\{1, 2, 3, 4, \ldots\}$ are used for counting, ordering and labelling and represented by the set $\mathbb{N}$. When zero is included, $\mathbb{N}^0$ or $\mathbb{N}_0$ is used:

$$\mathbb{N}^0 = \mathbb{N}_0 = \{0, 1, 2, \ldots\}.$$

Note that negative numbers are not included. Natural numbers are used to subscript a quantity to distinguish one element from another, e.g. $x_1, x_2, x_3, x_4, \ldots, x_n$.

2.3.2 Integers

Integers include the natural numbers and are represented by the set $\mathbb{Z}$:

$$\mathbb{Z} = \{\ldots, -2, -1, 0, 1, 2, 3, \ldots\}.$$

2.3.3 Rational Numbers

Any number that equals the quotient of one integer divided by another non-zero integer, is a *rational* number and represented by the set $\mathbb{Q}$. For example: $2, \sqrt{16}, 0.25$ are rational numbers because

$$2 = 4/2$$
$$\sqrt{16} = 4 = 8/2$$
$$0.25 = 1/4.$$

2.3.4 Irrational Numbers

An *irrational* number cannot be expressed as the quotient of two integers. Irrational numbers never terminate, nor contain repeated sequences of digits; consequently, they are always subject to a small error when stored within a computer. Examples are

$$\sqrt{2} = 1.414\ 213\ 56\ldots$$
$$\phi = 1.618\ 033\ 98\ldots (\text{golden section})$$
$$e = 2.718\ 281\ 82\ldots$$
$$\pi = 3.141\ 592\ 65\ldots$$

2.3.5 Real Numbers

Rational and irrational numbers comprise the set of *real* numbers $\mathbb{R}$. Examples are 1.5, 0.004, 12.999 and 23.0.

2.3.6 Algebraic and Transcendental Numbers

Polynomial equations with rational coefficients have the form:

$$f(x) = ax^n + bx^{n-1} + cx^{n-2} + \cdots + C$$

such as

$$y = 3x^2 + 2x - 1$$

and their roots belong to the set of *algebraic* numbers $\mathbb{A}$. A consequence of this definition implies that all rational numbers are algebraic, since if

$$x = \frac{p}{q}$$

then

$$qx - p = 0$$

which is a polynomial. Numbers that are not roots to polynomial equations are *transcendental* numbers and include most irrational numbers, but not $\sqrt{2}$, since if

$$x = \sqrt{2}$$

then

$$x^2 - 2 = 0$$

which is a polynomial.

2.3.7 Imaginary Numbers

Imaginary numbers take the form $\pm bi$, and belong to the set $\mathbb{I}$, where

$$bi \in \mathbb{I}, \quad b \in \mathbb{R}, \quad i^2 = -1.$$

Although some mathematicians place i before its multiplier: $i4$, others place it after the multiplier: $4i$, which is the convention I use in this book. However, when i is associated with trigonometric functions, it is good practice to place it before the function to avoid any confusion with the function's angle. For example, $\sin \alpha i$ can imply that the angle is imaginary, which is possible, whereas $i \sin \alpha$ implies that the value of $\sin \alpha$ is imaginary, which is also possible. Consequently, parentheses are used to clarify constructs such as $\sin(\alpha i)$.

Imaginary numbers obey all the axioms associated with real numbers.

2.3.8 Complex Numbers

A *complex* number has a real and an imaginary part $a + bi$, either of which may be zero, and belong to the set $\mathbb{C}$, where

$$(a + bi) \in \mathbb{C}, \quad a, b \in \mathbb{R}, \quad i^2 = -1.$$

The following are all complex numbers

$$3.5, \quad 3 + 4i, \quad -4 - 6i, \quad 7i, \quad 5.5 + 6.7i.$$

A real number is a complex number – it just has no imaginary part. This leads to the idea that real and imaginary numbers are subsets of complex numbers:

$$\mathbb{R} \subset \mathbb{C}, \quad \mathbb{I} \subset \mathbb{C}.$$

Therefore, a complex number can be constructed in all sorts of ways

$$\sin \alpha + i \cos \beta, \quad 2 - i \tan \alpha, \quad 23 + y^2 i.$$

2.4 Representing Complex Numbers

This section explores various ways of representing complex numbers numerically and graphically.

2.4.1 Real and Imaginary Parts

The real and imaginary parts of a complex number z are isolated by the $\mathrm{Re}(z)$ and $\mathrm{Im}(z)$ functions. For example:

$$z = a + bi$$
$$a = \mathrm{Re}(z)$$
$$b = \mathrm{Im}(z).$$

These two functions permit one to construct formal algebraic definitions such as defining one complex number being equal to another. In words, one would say "two complex numbers are equal iff (*if and only if*) they have identical real **and** imaginary parts". e.g. given $z_1 = x_1 + y_1 i$ and $z_2 = x_2 + y_2 i$, then $z_1 = z_2$ iff $x_1 = x_2$ **and** $y_1 = y_2$. Using Re and Im, we can write:

$$z_1 = z_2 \quad \leftrightarrow \quad [\mathrm{Re}(z_1) = \mathrm{Re}(z_2) \wedge \mathrm{Im}(z_1) = \mathrm{Im}(z_2)].$$

2.4.2 The Complex Plane

When the real number line is combined with a vertical imaginary axis, it creates the *complex plane*, as shown in Fig. 2.1, where any complex number has a unique position. Figure 2.2 shows the positions of the following four complex numbers

$$P = 4 + 3i, \quad Q = -3 + 2i, \quad R = -3 - 3i, \quad S = 4 - 5i.$$

Fig. 2.1 The complex plane

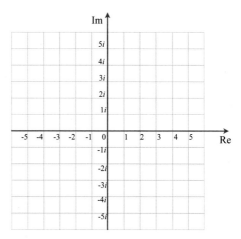

Fig. 2.2 Four complex numbers

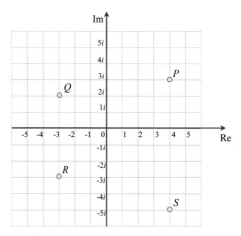

2.5 Complex Algebra

This section reviews the axioms associated with complex algebra, the complex conjugate, complex division, powers of i, the rotational properties of i, polar notation, the complex norm, the complex inverse, complex exponentials, the roots and logarithms of a complex number, hyperbolic functions, and the role of the complex plane in visualising complex functions.

2.5.1 Algebraic Laws

Complex numbers obey the axioms associated with real numbers. But for clarity, examples are included to show how imaginary terms are resolved.

Given

$$z_1 = x_1 + y_1 i$$
$$z_2 = x_2 + y_2 i$$
$$z_3 = x_3 + y_3 i.$$

The commutative law of addition:

$$z_1 + z_2 = z_2 + z_1 = (x_1 + x_2) + (y_1 + y_2)i$$
e.g. $(2 + 3i) + (4 + 5i) = 6 + 8i.$

The commutative law of multiplication:

$$z_1 z_2 = z_2 z_1$$
e.g. $(2 + 3i)(4 + 5i) = 8 + 10i + 12i + 15i^2$
$$= -7 + 22i.$$

The associative law of addition:

$$z_1 + (z_2 + z_3) = (z_1 + z_2) + z_3 = z_1 + z_2 + z_3$$
e.g. $(2 + 3i) + (4 + 5i) + (6 + 7i) = 12 + 15i.$

The associative law of multiplication:

$$z_1(z_2 z_3) = (z_1 z_2)z_3 = z_1 z_2 z_3$$
e.g. $(2 + 3i)(4 + 5i)(6 + 7i) = (8 + 22i + 15i^2)(6 + 7i)$
$$= (-7 + 22i)(6 + 7i)$$
$$= -42 + 132i - 49i + 154i^2$$
$$= -196 + 83i.$$

The distributive law of multiplication:

$$z_1(z_2 + z_3) = z_1 z_2 + z_1 z_3$$
e.g. $(2 + 3i) [(4 + 5i) + (6 + 7i)] = (2 + 3i)(10 + 12i)$
$$= 20 + 30i + 24i + 36i^2$$
$$= -16 + 54i.$$

From the above, one can see that the addition of complex numbers is identical to the addition of vectors. Figure 2.3 illustrates the addition of $z_1 = -2 + 3i$ and $z_2 = 3 + i.$

Fig. 2.3 The addition of $z_1 + z_2$

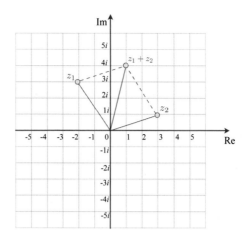

2.5.2 *Complex Conjugate*

The *complex conjugate* is a useful algebraic construct and is denoted by $\bar{z}$ or z^*. To avoid confusion, I will use $\bar{z}$ for complex numbers, $\bar{A}$ for conjugating a matrix, and A^* for the conjugate transpose of a matrix.

Given $z = a + bi$, then $\bar{z} = a - bi$. Also

$$\bar{z} = \text{Re}(z) - \text{Im}(z)i.$$

The product $z\bar{z}$ is extremely useful, as it is a real quantity. Generally,

$$
\begin{aligned}
z &= a + bi \\
\bar{z} &= a - bi \\
z\bar{z} &= (a + bi)(a - bi) \\
&= a^2 - abi + abi - b^2 i^2 \\
&= a^2 + b^2
\end{aligned}
$$

which is real. For example,

$$
\begin{aligned}
z &= 3 + 4i \\
\bar{z} &= 3 - 4i \\
z\bar{z} &= 25.
\end{aligned}
$$

Figure 2.4 shows z and $\bar{z}$.

Let's prove that $\overline{z_1 z_2} = \bar{z}_1 \bar{z}_2$. Given

Fig. 2.4 A complex number and its complex conjugate

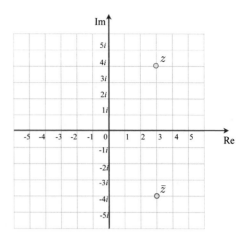

$$z_1 = a_1 + b_1 i$$
$$z_2 = a_2 + b_2 i$$
$$\overline{z_1 z_2} = \overline{(a_1 + b_1 i)(a_2 + b_2 i)}$$
$$= \overline{a_1 (a_2 + b_2 i)} + \overline{b_1 i (a_2 + b_2 i)}$$
$$= a_1 (a_2 - b_2 i) - b_1 i (a_2 - b_2 i)$$
$$= (a_1 - b_1 i)(a_2 - b_2 i)$$
$$= \overline{z_1} \, \overline{z_2}.$$

2.5.3 Complex Division

The complex conjugate is useful in resolving the quotient of two complex numbers; for if we multiply the numerator and the denominator by the complex conjugate of the denominator, the denominator becomes a real quantity and simplifies the division. For example, we evaluate this quotient as follows

$$z = \frac{10 + 5i}{1 + 2i}$$
$$= \frac{(10 + 5i)}{(1 + 2i)} \frac{(1 - 2i)}{(1 - 2i)}$$
$$= \frac{(10 + 5i)(1 - 2i)}{5}$$
$$= (2 + i)(1 - 2i)$$
$$= 2 - 4i + i - 2i^2$$
$$= 4 - 3i.$$

Table 2.1 Increasing powers of i

i^0	i^1	i^2	i^3	i^4	i^5	i^6
1	i	-1	$-i$	1	i	-1

2.5.4 Powers of i

As $i^2 = -1$, it must be possible to raise i to other powers. For example,

$$i^4 = i^2 i^2 = 1$$

and

$$i^5 = i i^4 = i.$$

Table 2.1 shows the sequence up to i^6.

This cyclic pattern is quite striking, and reminds one of:

$$(x, y, -x, -y, x, ...)$$

that arises when rotating around the Cartesian axes in a anti-clockwise direction. The above sequence is summarised as

$$\left. \begin{array}{l} i^{4n} = 1 \\ i^{4n+1} = i \\ i^{4n+2} = -1 \\ i^{4n+3} = -i \end{array} \right\} \quad \text{where } n \in \mathbb{N}^0.$$

But what about negative powers? Well they, too, are also possible. Consider i^{-1}, which is evaluated as follows

$$i^{-1} = \frac{1}{i} = \frac{1(-i)}{i(-i)} = \frac{-i}{1} = -i.$$

Similarly,

$$i^{-2} = \frac{1}{i^2} = \frac{1}{-1} = -1$$

and

$$i^{-3} = i^{-1} i^{-2} = -i(-1) = i.$$

Table 2.2 Decreasing powers of i

i^0	i^{-1}	i^{-2}	i^{-3}	i^{-4}	i^{-5}	i^{-6}
1	$-i$	-1	i	1	$-i$	-1

Table 2.2 shows the sequence down to i^{-6}.

This time the cyclic pattern is reversed and is similar to

$$(x, -y, -x, y, x, \ldots)$$

that arises when rotating around the Cartesian axes in a clockwise direction.

2.5.5 Rotational Qualities of i

Now let's investigate how a real number behaves when it is repeatedly multiplied by i. Starting with the number 3, we have,

$$i \times 3 = 3i$$
$$i \times 3i = -3$$
$$i \times (-3) = -3i$$
$$i \times (-3)i = 3.$$

The cycle is $(3, 3i, -3, -3i, 3, 3i, -3, -3i, 3, \ldots)$, which has four steps, as shown in Fig. 2.5.

Fig. 2.5 The cycle of points created by repeatedly multiplying 3 by i

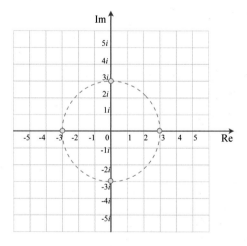

Fig. 2.6 Multiplying a
complex number by i

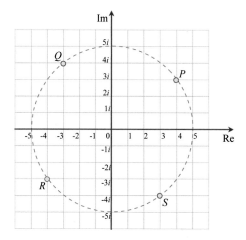

If we multiply a complex number by i, it is also rotated 90°. For example, the
complex number $P = 4 + 3i$ in Fig. 2.6 is rotated 90° to Q by multiplying it by i,

$$i(4 + 3i) = 4i + 3i^2$$
$$= 4i - 3$$
$$= -3 + 4i.$$

The point $Q = -3 + 4i$ is rotated 90° to R by multiplying it by i,

$$i(-3 + 4i) = -3i + 4i^2$$
$$= -3i - 4$$
$$= -4 - 3i.$$

The point $R = -4 - 3i$ is rotated 90° to S by multiplying it by i,

$$i(-4 - 3i) = -4i - 3i^2$$
$$= -4i + 3$$
$$= 3 - 4i.$$

Finally, the point $S = 3 - 4i$ is rotated 90° back to P by multiplying it by i,

$$i(3 - 4i) = 3i - 4i^2$$
$$= 3i + 4$$
$$= 4 + 3i.$$

As you can see, complex numbers are intimately related to Cartesian coordinates, in that the ordered pair $(x, \ y) \equiv (a + bi)$.

2.5.6 Modulus and Argument

As a complex number has a unique position on the complex plane, and is always relative to the origin of the real and imaginary axes, it can be visualised as a position vector and assigned a *modulus* or *magnitude*, which is the distance r of the complex point to the origin; consequently, $z = a + bi$ has a modulus $r = \sqrt{a^2 + b^2}$ and notated as $|z| = \sqrt{a^2 + b^2}$. This can also be expressed as

$$|z|^2 = a^2 + b^2 = [\operatorname{Re}(z)]^2 + [\operatorname{Im}(z)]^2 .$$

Here are some useful relationships:

$$z\bar{z} = (a + bi)(a - bi) = a^2 + b^2$$
$$z\bar{z} = |z|^2$$
$$|z| = |\bar{z}|$$
$$|-z| = |z|$$
$$|z_1 z_2| = |z_1||z_2|$$
$$|z_1 z_2|^2 = |z_1|^2|z_2|^2$$

Pursuing the similarity between complex numbers and position vectors, the straight line from the origin to a complex number $z = a + bi$, subtends with the real axis an angle θ, called the *argument*; consequently, $\theta = \tan^{-1}(b/a)$, and is notated as $\arg(z) = \tan^{-1}(b/a)$. Figure 2.7 shows the complex number $z = 3 + 4i$ with a modulus $r = \sqrt{3^2 + 4^2} = 5$ and an argument $\theta = \tan^{-1}(4/3) \approx 53.125°$.

From Fig. 2.7 we can generalise that a complex number $z = a + bi$ has real and imaginary components,

$$a = r \cos \theta$$
$$b = r \sin \theta$$

which permits us to state

$$z = r \cos \theta + ir \sin \theta$$

and when $r = 1$,

$$z = \cos \theta + i \sin \theta.$$

Fig. 2.7 The argument θ and modulus r of a complex number

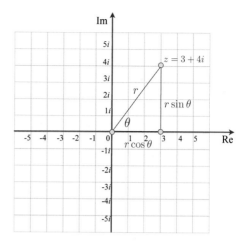

For example, given two complex numbers,

$$z_1 = a_1 + b_1 i$$
$$z_2 = a_2 + b_2 i$$

where

$$a_1 = r_1 \cos \theta_1, \quad b_1 = r_1 \sin \theta_1$$
$$a_2 = r_2 \cos \theta_2, \quad b_2 = r_2 \sin \theta_2$$

then

$$
\begin{aligned}
z_1 z_2 &= (a_1 + b_1 i)(a_2 + b_2 i) \\
&= (a_1 a_2 - b_1 b_2) + (a_1 b_2 + b_1 a_2)i \quad\quad (2.1)\\
&= (r_1 \cos \theta_1 r_2 \cos \theta_2 - r_1 \sin \theta_1 r_2 \sin \theta_2) \\
&\quad + (r_1 \cos \theta_1 r_2 \sin \theta_2 + r_1 \sin \theta_1 r_2 \cos \theta_2)i \\
&= r_1 r_2 (\cos \theta_1 \cos \theta_2 \sin \theta_1 \sin \theta_2) + i r_1 r_2 (\cos \theta_1 \sin \theta_2 + \sin \theta_1 \cos \theta_2) \\
&= r_1 r_2 \cos(\theta_1 + \theta_2) + i r_1 r_2 \sin(\theta_1 + \theta_2) \quad\quad (2.2)
\end{aligned}
$$

which shows that to compute the product of two complex numbers, we multiply their moduli and add their arguments. Let's illustrate this operation with an example. We'll start by computing the product using (2.1).

Given

$$z_1 = \tfrac{1}{2} + \tfrac{\sqrt{3}}{2}i$$

$$z_2 = -\tfrac{1}{2} + \tfrac{\sqrt{3}}{2}i$$

$$z_1 z_2 = \tfrac{1}{2}\left(-\tfrac{1}{2}\right) - \tfrac{\sqrt{3}}{2}\tfrac{\sqrt{3}}{2} + \left[\tfrac{1}{2}\tfrac{\sqrt{3}}{2} + \tfrac{\sqrt{3}}{2}\left(-\tfrac{1}{2}\right)\right]i$$

$$= -\tfrac{1}{4} - \tfrac{3}{4} + \left(\tfrac{\sqrt{3}}{4} - \tfrac{\sqrt{3}}{4}\right)i$$

$$= -1.$$

Next using (2.2). But first, we need to compute the moduli and arguments:

$$r_1 = \sqrt{\left(\tfrac{1}{2}\right)^2 + \left(\tfrac{\sqrt{3}}{2}\right)^2} = 1$$

$$r_2 = \sqrt{\left(-\tfrac{1}{2}\right)^2 + \left(\tfrac{\sqrt{3}}{2}\right)^2} = 1$$

$$\theta_1 = \tan^{-1}\left(\tfrac{\sqrt{3}}{2}\tfrac{2}{1}\right) = 60°$$

$$\theta_2 = \tan^{-1}\left(-\tfrac{\sqrt{3}}{2}\tfrac{2}{1}\right) = 120°$$

$$z_1 z_2 = \cos(60° + 120°) + i\sin(60° + 120°)$$

$$= -1.$$

Naturally, the results are the same.

2.5.7 Complex Norm

The term *norm* causes a lot of confusion, simply because there are so many, and each one requires a precise definition. For our purposes, norms are associated with vector spaces, where the norm of a vector is a function that returns some numerical property of the vector. The Euclidean norm of vector $\mathbf{v}$, is generally written

$$||\mathbf{v}|| = \sqrt{\mathbf{v} \cdot \mathbf{v}}$$

which is the square-root of the inner product of the vector with itself. For example, if vector $\mathbf{v} = [3 \quad 4]$, then $||\mathbf{v}|| = \sqrt{3^2 + 4^2} = 5$, and represents the Euclidean length of the vector.

The absolute value of a signed number $\pm x$ is written $|x|$. For example, if $x = +23$, $|x| = 23$, and if $x = -23$, $|x| = 23$. The absolute-value norm $||x||$, equals the absolute value, i.e. $||x|| = |x|$.

The Euclidean norm of a complex number $z = a + bi$ is given by

$$\|z\| = |z| = \sqrt{a^2 + b^2}$$

which is the modulus of z.

The modulus or Euclidean norm of a complex number measures an abstract distance corresponding to the length of the complex point to the origin on the complex plane, and is normally expressed:

$$\|z\| = \sqrt{z\bar{z}} = \sqrt{(a + bi)(a - bi)} = \sqrt{a^2 + b^2}.$$

Let's prove that the Euclidean norm of the product of two complex numbers, equals the product of the individual Euclidean norms.

$$z_1 = a_1 + b_1 i = r_1, \theta_1$$
$$z_2 = a_2 + b_2 i = r_2, \theta_2$$
$$\|z_1 z_2\| = |z_1 z_2|$$
$$= |z_1| \cdot |z_2|$$
$$= \|z_1\| \cdot \|z_2\|.$$

2.5.8 Complex Inverse

We have already seen that to divide a complex number x by another z, we multiply the numerator and denominator by the conjugate of the denominator:

$$\frac{x}{z} = \frac{a + bi}{c + di} \frac{c - di}{c - di}$$

which can be written as

$$xz^{-1} = x \frac{\bar{z}}{|z|^2}$$

thus the inverse of a complex number is

$$z^{-1} = \frac{\bar{z}}{|z|^2}.$$

For example, the inverse of $3 + 4i$ is

$$(3 + 4i)^{-1} = \tfrac{1}{25}(3 - 4i).$$

2.5.9   Complex Exponentials

In order to describe complex exponentials we require three power series. We start
with the power series for e^θ, $\sin\theta$ and $\cos\theta$,

$$e^\theta = 1 + \frac{\theta^1}{1!} + \frac{\theta^2}{2!} + \frac{\theta^3}{3!} + \frac{\theta^4}{4!} + \frac{\theta^5}{5!} + \frac{\theta^6}{6!} + \frac{\theta^7}{7!} + \frac{\theta^8}{8!} + \frac{\theta^9}{9!} + \cdots$$

$$\sin\theta = \theta - \frac{\theta^3}{3!} + \frac{\theta^5}{5!} - \frac{\theta^7}{7!} + \frac{\theta^9}{9!} + \cdots$$

$$\cos\theta = 1 - \frac{\theta^2}{2!} + \frac{\theta^4}{4!} - \frac{\theta^6}{6!} + \frac{\theta^8}{8!} + \cdots.$$

Euler discovered that by making θ imaginary: $e^{i\theta}$, we have

$$
\begin{aligned}
e^{i\theta} &= 1 + \frac{i\theta^1}{1!} - \frac{\theta^2}{2!} - \frac{i\theta^3}{3!} + \frac{\theta^4}{4!} + \frac{i\theta^5}{5!} - \frac{\theta^6}{6!} - \frac{i\theta^7}{7!} + \frac{\theta^8}{8!} + \frac{i\theta^9}{9!} \cdots \\
&= 1 - \frac{\theta^2}{2!} + \frac{\theta^4}{4!} - \frac{\theta^6}{6!} + \frac{\theta^8}{8!} + \cdots + \frac{i\theta^1}{1!} - \frac{i\theta^3}{3!} + \frac{i\theta^5}{5!} - \frac{i\theta^7}{7!} + \frac{i\theta^9}{9!} + \cdots \\
&= 1 - \frac{\theta^2}{2!} + \frac{\theta^4}{4!} - \frac{\theta^6}{6!} + \frac{\theta^8}{8!} + \cdots + i\left(\frac{\theta^1}{1!} - \frac{\theta^3}{3!} + \frac{\theta^5}{5!} - \frac{\theta^7}{7!} + \frac{\theta^9}{9!} + \cdots\right) \\
&= \cos\theta + i\sin\theta
\end{aligned}
$$

which is *Euler's trigonometric formula*. If we now reverse the sign of $i\theta$ to $-i\theta$, we
have

$$
\begin{aligned}
e^{-i\theta} &= 1 - \frac{i\theta^1}{1!} - \frac{\theta^2}{2!} + \frac{i\theta^3}{3!} + \frac{\theta^4}{4!} - \frac{i\theta^5}{5!} - \frac{\theta^6}{6!} + \frac{i\theta^7}{7!} + \frac{\theta^8}{8!} - \frac{i\theta^9}{9!} \cdots \\
&= 1 - \frac{\theta^2}{2!} + \frac{\theta^4}{4!} - \frac{\theta^6}{6!} + \frac{\theta^8}{8!} + \cdots - \frac{i\theta^1}{1!} + \frac{i\theta^3}{3!} - \frac{i\theta^5}{5!} + \frac{i\theta^7}{7!} + \frac{i\theta^9}{9!} + \cdots \\
&= 1 - \frac{\theta^2}{2!} + \frac{\theta^4}{4!} - \frac{\theta^6}{6!} + \frac{\theta^8}{8!} + \cdots - i\left(\frac{\theta^1}{1!} - \frac{\theta^3}{3!} + \frac{\theta^5}{5!} - \frac{\theta^7}{7!} + \frac{\theta^9}{9!} + \cdots\right) \\
&= \cos\theta - i\sin\theta
\end{aligned}
$$

thus we have

$$e^{i\theta} = \cos\theta + i\sin\theta$$
$$e^{-i\theta} = \cos\theta - i\sin\theta$$

from which we obtain

$$\cos\theta = \frac{e^{i\theta} + e^{-i\theta}}{2}$$

$$\sin\theta = \frac{e^{i\theta} - e^{-i\theta}}{2i}.$$

Given $e^{i\theta} = \cos\theta + i\sin\theta$, when $\theta = \pi$, we have $e^{i\pi} = -1$, or $e^{i\pi} + 1 = 0$, which is Euler's famous equation. The American physicist Richard Feynman (1918–1988) referred to the equation as "our jewel" and "the most remarkable formula in mathematics." [1]

Another strange formula emerges as follows:

$$\cos\theta + i\sin\theta = e^{i\theta}$$

$$\cos\left(\tfrac{\pi}{2}\right) + i\sin\left(\tfrac{\pi}{2}\right) = e^{i\pi/2}$$

$$i = e^{i\pi/2}$$

$$i^i = \left(e^{i\pi/2}\right)^i$$

$$= e^{i^2\pi/2}$$

$$= e^{-\pi/2}$$

$$i^i = 0.207\,879\,576\ldots$$

which reveals that i^i is a real number, even though i is not a number, as we know it!

Geometrically, $e^{i\theta}$ is a point on the unit circle, on the complex plane. Consequently, $re^{i\beta}$ is another point, radius r from the origin, with real and imaginary coordinates $x = r\cos\beta$ and $y = r\sin\beta$, respectively, as shown in Fig. 2.8. This is the polar form of a complex number.

Let's return to the product of two complex numbers, and see how the product can be visualised using polar notation.

Fig. 2.8 The unit circle and $re^{i\beta}$

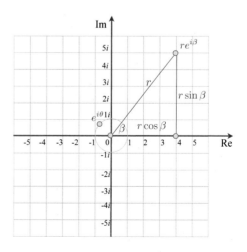

Fig. 2.9 Three complex
numbers

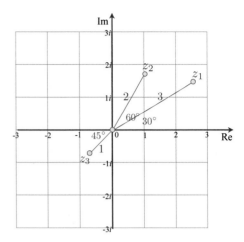

Equation (2.2) shows that

$$z_1 = r_1(\cos\theta_1 + i\sin\theta_1)$$
$$z_2 = r_2(\cos\theta_2 + i\sin\theta_2)$$
$$z_1 z_2 = r_1 r_2 \left(\cos(\theta_1 + \theta_2) + i\sin(\theta_1 + \theta_2)\right).$$

Using polar notation:

$$z_1 = r_1 e^{i\theta_1}$$
$$z_2 = r_2 e^{i\theta_2}$$
$$z_1 z_2 = r_1 e^{i\theta_1} r_2 e^{i\theta_2}$$
$$= r_1 r_2 e^{i(\theta_1 + \theta_2)}.$$

Now let's show three ways of computing the product of the following three complex
numbers shown in Fig. 2.9:

$$z_1 = \tfrac{3\sqrt{3}}{2} + \tfrac{3}{2}i$$
$$z_2 = 1 + \sqrt{3}i$$
$$z_3 = -\tfrac{\sqrt{2}}{2} - \tfrac{\sqrt{2}}{2}i.$$

z_1: the argument is 30° and a modulus of 3.
z_2: the argument is 60° and a modulus of 2.
z_3: the argument is 225° and a modulus of 1.
Let's compute the product $z_1 z_2 z_3$

$$z_1 z_2 z_3 = \left(\tfrac{3\sqrt{3}}{2} + \tfrac{3}{2}i \right) \left(1 + \sqrt{3}i \right) \left(-\tfrac{\sqrt{2}}{2} - \tfrac{\sqrt{2}}{2}i \right)$$

$$= \left(\tfrac{3\sqrt{3}}{2} + \tfrac{9}{2}i + \tfrac{3}{2}i - \tfrac{3\sqrt{3}}{2} \right) \left(-\tfrac{\sqrt{2}}{2} - \tfrac{\sqrt{2}}{2}i \right)$$

$$= 6i \left(-\tfrac{\sqrt{2}}{2} - \tfrac{\sqrt{2}}{2}i \right)$$

$$= 3\sqrt{2} - 3\sqrt{2}i$$

which confirms that the product $z_1 z_2 z_3$ rotates any complex number 315°, and scales its modulus by 6.

Now let's compute the product using cosines and sines.

$$z_1 = 3(\cos 30° + i \sin 30°)$$

$$z_2 = 2(\cos 60° + i \sin 60°)$$

$$z_3 = \cos 225° + i \sin 225°$$

$$z_1 z_2 z_3 = 3(\cos 30° + i \sin 30°)2(\cos 60° + i \sin 60°)(\cos 225° + i \sin 225°)$$

$$= 6(\cos 315° + i \sin 315°)$$

$$= 3\sqrt{2} - 3\sqrt{2}i$$

which is much simpler. Finally, let's define the complex numbers in polar form, with angles in degrees, for clarity.

$$z_1 = 3e^{i30°}$$

$$z_2 = 2e^{i60°}$$

$$z_3 = e^{i225°}$$

$$z_1 z_2 z_3 = 3e^{i30°} 2e^{i60°} e^{i225°}$$

$$= 6e^{315°}$$

$$= 6(\cos 315° + i \sin 315°)$$

$$= 3\sqrt{2} - 3\sqrt{2}i$$

which is even neater!

2.5.10 de Moivre's Theorem

Euler's trigonometric formula can be developed as follows.

$$\cos\theta + i \sin\theta = e^{i\theta}$$

$$(\cos\theta + i \sin\theta)^n = \left(e^{i\theta} \right)^n$$

$$= e^{in\theta}$$

$$(\cos\theta + i \sin\theta)^n = \cos(n\theta) + i \sin(n\theta) \qquad (2.3)$$

where (2.3) is known as *de Moivre's theorem*, after Abraham de Moivre.

Substituting $n = 2$ in (2.3) we obtain

$$\cos(2\theta) + i\sin(2\theta) = (\cos\theta + i\sin\theta)^2$$
$$= \cos^2\theta - \sin^2\theta + 2i\cos\theta\sin\theta.$$

Therefore,

$$\cos(2\theta) = \mathrm{Re}\left(\cos^2\theta - \sin^2\theta + 2i\cos\theta\sin\theta\right)$$
$$= \cos^2\theta - \sin^2\theta$$
$$\sin(2\theta) = \mathrm{Im}\left(\cos^2\theta - \sin^2\theta + 2i\cos\theta\sin\theta\right)$$
$$= 2\cos\theta\sin\theta.$$

de Moivre's theorem can be used for similar identities by substituting other values of n. Let's try $n = 3$:

$$\cos(3\theta) + i\sin(3\theta) = (\cos\theta + i\sin\theta)^3$$
$$= (\cos\theta + i\sin\theta)(\cos^2\theta - \sin^2\theta + 2i\cos\theta\sin\theta).$$

Therefore,

$$\cos(3\theta) = \mathrm{Re}\left[(\cos\theta + i\sin\theta)(\cos^2\theta - \sin^2\theta + 2i\cos\theta\sin\theta)\right]$$
$$= \cos^3\theta - \cos\theta\sin^2\theta - 2\cos\theta\sin^2\theta$$
$$= \cos^3\theta - 3\cos\theta\sin^2\theta$$
$$= \cos^3\theta - 3\cos\theta(1 - \cos^2\theta)$$
$$= 4\cos^3\theta - 3\cos\theta.$$
$$\sin(3\theta) = \mathrm{Im}\left[(\cos\theta + i\sin\theta)(\cos^2\theta - \sin^2\theta + 2i\cos\theta\sin\theta)\right]$$
$$= \cos^2\theta\sin\theta - \sin^3\theta + 2\cos^2\theta\sin\theta$$
$$= 3\cos^2\theta\sin\theta - \sin^3\theta$$
$$= 3\sin\theta(1 - \sin^2\theta) - \sin^3\theta$$
$$= 3\sin\theta - 4\sin^3\theta.$$

Let's test these identities with $\theta = 30°$:

$$\cos(3\theta) = 4\cos^3\theta - 3\cos\theta$$
$$\cos 90° = 4\cos^3 30° - 3\cos 30°$$
$$= 4\left(\tfrac{\sqrt{3}}{2}\right)^3 - 3\tfrac{\sqrt{3}}{2}$$
$$= \tfrac{3}{2}\sqrt{3} - \tfrac{3}{2}\sqrt{3}$$

$$= 0.$$
$$\sin(3\theta) = 3\sin\theta - 4\sin^3\theta$$
$$\sin 90° = 3\sin 30° - 4\sin^3 30°$$
$$= \tfrac{3}{2} - 4\left(\tfrac{1}{2}\right)^3$$
$$= 1.$$

Given $z = \cos\theta + i\sin\theta$, we can define $\cos\theta$ and $\sin\theta$ in terms of z as follows.

$$z = \cos\theta + i\sin\theta \tag{2.4}$$
$$= e^{i\theta}$$
$$\tfrac{1}{z} = e^{-i\theta}$$
$$= \cos\theta - i\sin\theta \tag{2.5}$$

adding and subtracting (2.4) and (2.5):

$$z + \tfrac{1}{z} = 2\cos\theta$$
$$z - \tfrac{1}{z} = 2i\sin\theta$$
$$\cos\theta = \tfrac{1}{2}\left(z + \tfrac{1}{z}\right) \tag{2.6}$$
$$\sin\theta = \tfrac{-i}{2}\left(z - \tfrac{1}{z}\right). \tag{2.7}$$

Let's use de Moivre's formula to show that

$$z^n + \frac{1}{z^n} = 2\cos(n\theta).$$

Proof:

$$z^n = \cos(n\theta) + i\sin(n\theta)$$
$$z^{-n} = \cos(-n\theta) + i\sin(-n\theta)$$
$$= \cos(n\theta) - i\sin(n\theta)$$
$$z^n + z^{-n} = 2\cos(n\theta).$$

2.5.11 nth Root of Unity

The real roots of 1 can only be ± 1, but complex numbers introduce the concept that unity possesses an infinite number of complex roots. The complex roots of 1 satisfy the equation $z^n = 1$, where n is a positive integer. Such roots are are employed in different branches of mathematics, such as number theory and discrete Fourier transforms. (See https://en.wikipedia.org/wiki/Root_of_unity)

If the nth root of 1 is z, then $z^n = 1$. Therefore, using de Moivre's theorem:

$$1^{1/n} = e^{i2\pi k/n}, \quad k = 0, 1, 2, \ldots, n-1$$
$$= \cos\left(\frac{2\pi k}{n}\right) + i \sin\left(\frac{2\pi k}{n}\right).$$

For example, when $n = 3$:

$$[k = 0] \quad z_0 = \cos\left(\frac{0}{3}\right) + i \sin\left(\frac{0}{3}\right) = 1$$
$$[k = 1] \quad z_1 = \cos\left(\frac{2\pi}{3}\right) + i \sin\left(\frac{2\pi}{3}\right) = -\frac{1}{2} + i\frac{\sqrt{3}}{2}$$
$$[k = 2] \quad z_1 = \cos\left(\frac{4\pi}{3}\right) + i \sin\left(\frac{4\pi}{3}\right) = -\frac{1}{2} - i\frac{\sqrt{3}}{2}.$$

Let's confirm these results:

$$z_1^3 = \left(-\frac{1}{2} + i\frac{\sqrt{3}}{2}\right)\left(-\frac{1}{2} + i\frac{\sqrt{3}}{2}\right)\left(-\frac{1}{2} + i\frac{\sqrt{3}}{2}\right)$$
$$= \left(-\frac{1}{2} - i\frac{\sqrt{3}}{2}\right)\left(-\frac{1}{2} + i\frac{\sqrt{3}}{2}\right) = 1$$
$$z_2^3 = \left(-\frac{1}{2} - i\frac{\sqrt{3}}{2}\right)\left(-\frac{1}{2} - i\frac{\sqrt{3}}{2}\right)\left(-\frac{1}{2} - i\frac{\sqrt{3}}{2}\right)$$
$$= \left(-\frac{1}{2} + i\frac{\sqrt{3}}{2}\right)\left(-\frac{1}{2} - i\frac{\sqrt{3}}{2}\right) = 1.$$

These roots are located on the unit-radius complex circle, as shown in Fig. 2.10. Connecting the points together creates a regular polygon.

Fig. 2.10 Three roots of unity

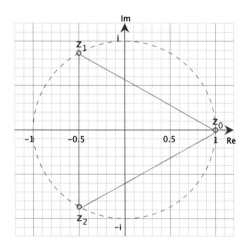

2.5.12 nth Roots of a Complex Number

Given

$$z = r(\cos\theta + i\sin\theta)$$

then

$$\sqrt[n]{z} = \sqrt[n]{r}\left[\cos\left(\tfrac{\theta+k2\pi}{n}\right) + i\sin\left(\tfrac{\theta+k2\pi}{n}\right)\right], \quad 0 \le k \le (n-1).$$

For example, let's find $\left(8 + i8\sqrt{3}\right)^{1/4}$.

To begin, we convert it into polar form: $z = 16e^{i\pi/3}$.

$$z = 16\left[\cos\left(\tfrac{\pi}{3}\right) + i\sin\left(\tfrac{\pi}{3}\right)\right]$$

$$\sqrt[4]{z} = \sqrt[4]{16}\left[\cos\left(\tfrac{\pi/3+k2\pi}{4}\right) + i\sin\left(\tfrac{\pi/3+k2\pi}{4}\right)\right]$$

$$[k=0] \quad z_0 = 2\left[\cos\left(\tfrac{\pi}{12}\right) + i\sin\left(\tfrac{\pi}{12}\right)\right] \approx 1.932 + i0.518$$

$$[k=1] \quad z_1 = 2\left[\cos\left(\tfrac{\pi}{12} + \tfrac{2\pi}{4}\right) + i\sin\left(\tfrac{\pi}{12} + \tfrac{2\pi}{4}\right)\right] \approx -0.518 + i1.932$$

$$[k=2] \quad z_2 = 2\left[\cos\left(\tfrac{\pi}{12} + \tfrac{4\pi}{4}\right) + i\sin\left(\tfrac{\pi}{12} + \tfrac{4\pi}{4}\right)\right] \approx -1.932 - i0.518$$

$$[k=3] \quad z_3 = 2\left[\cos\left(\tfrac{\pi}{12} + \tfrac{6\pi}{4}\right) + i\sin\left(\tfrac{\pi}{12} + \tfrac{6\pi}{4}\right)\right] \approx 0.518 - i1.932.$$

Figure 2.11 shows these roots.

Fig. 2.11 4th roots of $16(\cos(\pi/3) + i\sin(\pi/3))$

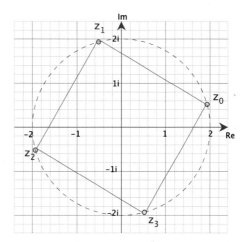

2.5.13 Logarithm of a Complex Number

In order to take the natural logarithm of a complex number, we use the exponential form. Consequently, if we are given $a + bi$, this must be converted to $re^{i\theta}$. Therefore, given

$$z = re^{i\theta}$$

then

$$\ln z = \ln r + i\theta$$

or

$$\ln z = \ln |z| + i \arg(z).$$

As exponential functions can have multiple values, the imaginary component is restricted to the interval $-\pi < \theta \leq \pi$. For example, -1 is represented by $e^{i\pi}$, $e^{3i\pi}$, $e^{5i\pi}$ etc., but to satisfy the agreed interval constraint, $-1 = e^{i\pi}$.

For example, given $z = -2 + 2i$, then

$$-2 + 2i = \sqrt{(-2)^2 + 2^2} \cdot e^{i \tan^{-1}(2/-2)}$$
$$= \sqrt{8}e^{i0.75\pi}$$
$$\ln(-2 + 2i) = \ln\left(\sqrt{8}e^{i0.75\pi}\right)$$
$$= \ln\sqrt{8} + 0.75\pi i$$
$$\approx 1.039721 + 2.356194i.$$

Similarly, given $z = 3 - 4i$, then

$$3 - 4i = \sqrt{3^2 + (-4)^2} \cdot e^{i \tan^{-1}(-4/3)}$$
$$= 5e^{-i0.927295}$$
$$\ln(3 - 4i) = \ln\left(5e^{-i0.927295}\right)$$
$$= \ln 5 - 0.927295i$$
$$\approx 1.609438 - 0.927295i.$$

Logarithms of other complex numbers are shown in Table 2.3.

Table 2.3 Logarithms of complex numbers

z	e form	$\ln z$
1	e^{i0}	0
-1	$e^{i\pi}$	πi
i	$e^{i\pi/2}$	$\frac{\pi}{2}i$
$-i$	$e^{-i\pi/2}$	$-\frac{\pi}{2}i$
5	$5e^{i0}$	1.609438
-5	$5e^{i\pi}$	$1.609438 + \pi i$
$5i$	$5e^{i\pi/2}$	$1.609438 + \frac{\pi}{2}i$
$-5i$	$5e^{-i\pi/2}$	$1.609438 - \frac{\pi}{2}i$
$5 + 5i$	$\sqrt{50}e^{i\pi/4}$	$1.956012 + \frac{\pi}{4}i$
$5 - 5i$	$\sqrt{50}e^{-i\pi/4}$	$1.956012 - \frac{\pi}{4}i$
$-5 + 5i$	$\sqrt{50}e^{i3\pi/4}$	$1.956012 + \frac{3\pi}{4}i$
$-5 - 5i$	$\sqrt{50}e^{-i3\pi/4}$	$1.956012 - \frac{3\pi}{4}i$
0.5	$0.5e^{i0}$	-0.693147
-0.5	$0.5e^{i\pi}$	$-0.693147 + \pi i$
$0.5i$	$0.5e^{\pi/2}$	$-0.693147 + \frac{\pi}{2}i$
$-0.5i$	$0.5e^{-i\pi/2}$	$-0.693147 - \frac{\pi}{2}i$

2.5.14   *Raising a Complex Number to a Complex Power*

Now that we have seen how to take a logarithm of a complex number, the way is open to raise a complex number to a complex power. For example, given

$$z = e^{y} \tag{2.8}$$

then

$$y = \ln z \tag{2.9}$$

and substituting (2.9) in (2.8), we obtain

$$z = e^{\ln z}. \tag{2.10}$$

Raising both sides of (2.10) to some power w,

$$z^{w} = \left(e^{\ln z}\right)^{w} = e^{w \ln z}. \tag{2.11}$$

Equation 2.11 applies to both real and complex numbers, so first, let's begin with

$$z = 2$$
$$w = 1 + i$$

which requires raising e to the product of $1 + i$ and the natural logarithm of 2.

$$\ln 2 \approx 0.693147$$
$$(1 + i)0.693147 = 0.693147 + 0.693147i$$
$$e^{(0.693147+0.693147i)} = e^{0.693147}e^{0.693147i}$$
$$= 2(\cos 0.693147 + i \sin 0.693147)$$
$$\approx 2(0.769239 + 0.638961i)$$
$$= 1.538478 + 1.277922i$$

therefore,

$$2^{1+i} \approx 1.538478 + 1.277922i.$$

Now let's use

$$z = 2 + 2i$$
$$w = 1 + i$$

then

$$z^w = (2 + 2i)^{1+i} = e^{(1+i)\ln(2+2i)}$$

which requires raising e to the product of $1 + i$ and the natural logarithm of $2 + 2i$. Not very nice, but let's have a go!

$$2 + 2i = \sqrt{2^2 + 2^2}e^{i\pi/4}$$
$$= \sqrt{8}e^{i\pi/4}$$
$$\ln(2 + 2i) = \ln 2.828427 + \frac{i\pi}{4}$$
$$\approx 1.039721 + 0.785398i$$
$$(1 + i)(1.039721 + 0.785398i) = 1.039721 + 1.039721i + 0.785398i - 0.785398$$
$$= 0.254323 + 1.825119i$$
$$e^{(0.254323+1.825119i)} = e^{0.254323}e^{1.825119i}$$
$$= 1.289588(\cos 1.825119 + i \sin 1.825119)$$
$$\approx 1.289588(-0.25159 + 0.967834i)$$
$$\approx -0.324447 + 1.248107i$$

therefore,

$$(2 + 2i)^{1+i} \approx -0.324447 + 1.248107i.$$

2.5.15 Visualising Simple Complex Functions

We are aware of how real functions such as $f(x) = 2x^2 + 3x + 5$ behave, as it is possible to draw a graph relating $f(x)$ to x. But when it comes to complex functions, such as $f(z) = (a + bi)^2$, we require two dimensions to represent the original real and imaginary terms, and two further dimensions to represent the function, which is difficult in our three-dimensional world. However, in order to get a feel for what is happening between a complex variable and function, we can plot how individual numbers behave when subject to a function. To illustrate this, Fig. 2.12 illustrates how nine complex numbers in the first quadrant, behave when they are subject to a square function. For example, $(1 + 3i)^2$ moves to $-8 + 6i$, and $(3 + 3i)^2$ moves to $18i$. The dashed lines show the trajectory as the exponent increases from 1 to 2. Note that the squaring function imposes an anti-clockwise rotation on the trajectories, with the end complex numbers in the same or second quadrant. The solid blue lines connect the transformed points together to emphasise the distortion caused by the squaring transformation.

The functions used to draw the blue lines are

$$\left.\begin{aligned}
f(z) &= [(1 + i)(1 - t) + (1 + 3i)t]^2 \\
f(z) &= [(2 + i)(1 - t) + (2 + 3i)t]^2 \\
f(z) &= [(3 + i)(1 - t) + (3 + 3i)t]^2 \\
f(z) &= [(1 + i)(1 - t) + (3 + i)t]^2 \\
f(z) &= [(1 + 2i)(1 - t) + (3 + 2i)t]^2 \\
f(z) &= [(1 + 3i)(1 - t) + (3 + 3i)t]^2
\end{aligned}\right\} \quad 0 \le t \le 1.$$

Fig. 2.12 The trajectories of nine complex numbers in the first quadrant, when squared

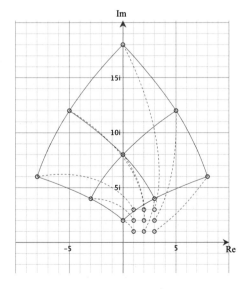

Fig. 2.13 The trajectories of nine complex numbers in the second quadrant, when squared

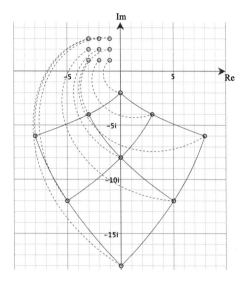

Fig. 2.14 The trajectories of nine complex numbers in the third quadrant, when squared

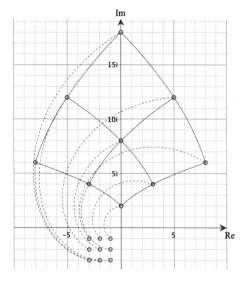

Figure 2.13 shows the trajectories for nine similar complex numbers in the second quadrant, where the squaring function imposes an anti-clockwise rotation on the trajectories, with the end complex numbers in the third or fourth quadrant.

Figure 2.14 shows the trajectories for nine complex numbers in the third quadrant, where the squaring function imposes a clockwise rotation on the trajectories, with the end complex numbers in the first or second quadrant.

Figure 2.15 shows the trajectories for nine complex numbers in the fourth quadrant, where the squaring function imposes a clockwise rotation on the trajectories, with the end complex numbers in the third or fourth quadrant.

Fig. 2.15 The trajectories of
nine complex numbers in the
fourth quadrant, when
squared

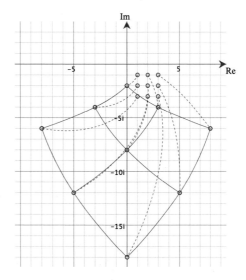

Fig. 2.16 The trajectories of
eight, squared imaginary
numbers

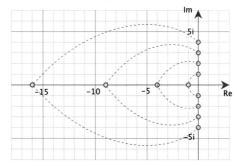

The only remaining numbers to consider are on the real and imaginary axes. The
real axis is simple, as the square of any real number is another real number. Figure 2.16
shows the anti-clockwise trajectories of four positive imaginary numbers, and the
clockwise trajectories of four negative imaginary numbers.

2.5.16 The Hyperbolic Functions

The trigonometric functions derive from the geometry of the circle $x^2 + y^2 = 1$,
whereas the *hyperbolic functions* are associated with the geometry of the hyperbola
$x^2 - y^2 = 1$. However, they are all related to e as we will see.
  Given

$$\cos\theta = \frac{e^{i\theta} + e^{-i\theta}}{2}$$

then

$$\cos(i\theta) = \frac{e^{i(i\theta)} + e^{-i(i\theta)}}{2} = \frac{e^{-\theta} + e^{\theta}}{2} = \cosh\theta.$$

Similarly, given

$$\sin\theta = \frac{e^{i\theta} - e^{-i\theta}}{2i}$$

then

$$\sin(i\theta) = \frac{e^{ii\theta} - e^{-ii\theta}}{2i} = \frac{e^{-\theta} - e^{\theta}}{2i} = \frac{i(e^{\theta} - e^{-\theta})}{2} = i\sinh\theta.$$

By definition:

$$\tanh\theta = \frac{\sinh\theta}{\cosh\theta} = \frac{e^{\theta} - e^{-\theta}}{e^{\theta} + e^{-\theta}}.$$

Therefore,

$$\cosh\theta = \frac{e^{\theta} + e^{-\theta}}{2} = \cos(i\theta)$$

$$\sinh\theta = \frac{e^{\theta} - e^{-\theta}}{2} = -i\sin(i\theta)$$

$$\tanh\theta = \frac{e^{\theta} - e^{-\theta}}{e^{\theta} + e^{-\theta}} = -i\tan(i\theta)$$

$$\cosh\theta + \sinh\theta = e^{\theta}$$

$$\cosh\theta - \sinh\theta = e^{-\theta}$$

$$\cosh^2\theta - \sinh^2\theta = 1.$$

Figure 2.17 shows how sinh and cosh relate to the hyperbola.

2.5.17   Derivative of a Complex Number

Basic calculus informs us that

$$\frac{d}{d\theta}e^{n\theta} = ne^{n\theta}$$

Fig. 2.17 The hyperbolic
functions

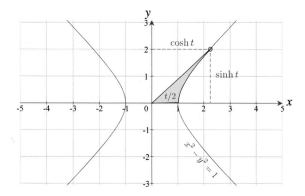

therefore,

$$\frac{d}{d\theta}e^{i\theta} = ie^{i\theta}$$

and

$$\frac{d}{d\theta}(\cos\theta + i\sin\theta) = -\sin\theta + i\cos\theta = i(\cos\theta + i\sin\theta).$$

We also expect the derivative of the exponential form to agree:

$$\begin{aligned}
\frac{d}{d\theta}\cos\theta &= \frac{d}{d\theta}\mathrm{Re}\left(e^{i\theta}\right) \\
&= \frac{d}{d\theta}\left[\tfrac{1}{2}\left(e^{i\theta} + e^{-i\theta}\right)\right] \\
&= \tfrac{i}{2}\left(e^{i\theta} - e^{-i\theta}\right) \\
&= -\sin\theta
\end{aligned}$$

$$\begin{aligned}
\frac{d}{d\theta}\sin\theta &= \frac{d}{d\theta}\mathrm{Im}\left(e^{i\theta}\right) \\
&= \frac{d}{d\theta}\left[\tfrac{1}{2i}\left(e^{i\theta} - e^{-i\theta}\right)\right] \\
&= \tfrac{1}{2}\left(e^{i\theta} + e^{-i\theta}\right) \\
&= \cos\theta.
\end{aligned}$$

2.6 Summary

Hopefully, this chapter has established $i = \sqrt{-1}$ as an incredible invention. Even though it does not belong to the traditional number systems, it is a valid mathematical object and reveals hidden numerical relationships between various constants and functions. Perhaps the two outstanding examples being $e^{i\pi} + 1 = 0$ and $i^i = 0.207\,879\ldots$.

The complex plane provides a simple way of visualising complex numbers, and illustrates their connection with vectors.

Euler's proof for $e^{i\theta} = \cos\theta + i\sin\theta$ opens the door for associating complex numbers with wave phenomena, which include acoustic waves, sea waves, electronics and quantum mechanics. These are covered in later chapters.

2.6.1 Summary of Complex Formulae

For all the following formulae, $z = a + bi$ and $i^2 = -1$.
Imaginary Number

$$bi, \quad bi \in \mathbb{I}, \quad b \in \mathbb{R}.$$

Complex Number

$$z = a \pm bi, \quad (a \pm bi) \in \mathbb{C}, \quad a, b \in \mathbb{R}$$
$$\mathbb{R} \subset \mathbb{C}, \quad \mathbb{I} \subset \mathbb{C}.$$

Real and Imaginary Parts

$$a = \mathrm{Re}(z), \quad b = \mathrm{Im}(z).$$

Complex Conjugate

$$\bar{z} = a - bi$$
$$z\bar{z} = a^2 + b^2$$
$$\overline{z_1 z_2} = \bar{z}_1 \bar{z}_2.$$

Powers of i

$$i^{4n} = 1, \quad i^{4n+1} = i, \quad i^{4n+2} = -1, \quad i^{4n+3} = -i.$$

Modulus and Argument

$$z = (r, \theta), \quad r = \sqrt{a^2 + b^2}, \quad \theta = \tan^{-1}(b/a)$$
$$a = r \cos \theta$$
$$b = r \sin \theta$$
$$z = r(\cos \theta + i \sin \theta)$$
$$z_1 = a_1 + b_1 i = (r_1, \theta_1)$$
$$z_2 = a_2 + b_2 i = (r_2, \theta_2)$$
$$z_1 z_2 = r_1 r_2 [\cos(\theta_1 + \theta_2) + i \sin(\theta_1 + \theta_2)].$$

Complex Norm

$$\|z\| = |z| = \sqrt{z\bar{z}} = \sqrt{a^2 + b^2}.$$

Complex Inverse

$$z^{-1} = \frac{\bar{z}}{|z|^2}.$$

Complex Exponential

$$e^{i\theta} = \cos \theta + i \sin \theta$$
$$e^{-i\theta} = \cos \theta - i \sin \theta$$
$$\cos \theta = \frac{e^{i\theta} + e^{-i\theta}}{2}$$
$$\sin \theta = \frac{e^{i\theta} - e^{-i\theta}}{2i}$$
$$e^{i\pi} + 1 = 0$$
$$i^i = 0.207\ 879\ 576\ \ldots.$$

de Moivre's Theorem

$$(\cos \theta + i \sin \theta)^n = \cos(n\theta) + i \sin(n\theta)$$
$$\cos(2\theta) = \cos^2 \theta - \sin^2 \theta$$
$$\sin(2\theta) = 2 \cos \theta \sin \theta$$
$$\cos(3\theta) = 4 \cos^3 \theta - 3 \cos \theta$$
$$\sin(3\theta) = 3 \sin \theta - 4 \sin^3 \theta$$
$$z^n + \frac{1}{z^n} = 2 \cos(n\theta).$$

*n*th **Root of Unity**

$$z^n = 1$$
$$z = \cos\left(\frac{k2\pi}{n}\right) + i \sin\left(\frac{k2\pi}{n}\right).$$

*n*th **Root of a Complex Number**

$$\sqrt[n]{z} = \sqrt[n]{r}\left[\cos\left(\frac{\theta+k2\pi}{n}\right) + i \sin\left(\frac{\theta+k2\pi}{n}\right)\right], \quad 0 \le k \le n-1.$$

Logarithm of a Complex Number

$$z = re^{i\theta}$$
$$\ln z = \ln r + i\theta$$
$$\ln z = \ln|z| + i \arg(z).$$

Raising a Complex Number to a Complex Power

$$z^w = e^{w \ln z}.$$

The Hyperbolic Functions

$$\cosh\theta = \frac{e^\theta + e^{-\theta}}{2} = \cos(i\theta)$$

$$\sinh\theta = \frac{e^\theta - e^{-\theta}}{2} = -i\sin(i\theta)$$

$$\tanh\theta = \frac{e^\theta - e^{-\theta}}{e^\theta + e^{-\theta}} = -i\tan(i\theta)$$

$$\cosh\theta + \sinh\theta = e^\theta$$
$$\cosh\theta - \sinh\theta = e^{-\theta}$$
$$\cosh^2\theta - \sinh^2\theta = 1.$$

Derivative of a Complex Number

$$\frac{d}{d\theta}e^{i\theta} = ie^{i\theta}.$$

2.7 Worked Examples

2.7.1 Complex Addition

Compute $(3 + 2i) + (2 + 2i) + (5 - 3i)$.
Solution: Collect up like terms.

$$(3 + 2i) + (2 + 2i) + (5 - 3i) = 10 + i.$$

2.7.2 Complex Products

Compute $(3 + 2i)(2 + 2i)(5 - 3i)$.
Solution: Expand algebraically and simplify.

$$\begin{aligned}
(3 + 2i)(2 + 2i)(5 - 3i) &= (3 + 2i)(10 - 6i + 10i + 6) \\
&= (3 + 2i)(16 + 4i) \\
&= 48 + 12i + 32i - 8 \\
&= 40 + 44i.
\end{aligned}$$

2.7.3 Complex Division

Compute $\frac{1}{(2+3i)(4-5i)}$.
Solution: Expand the denominator, then multiply top and bottom by the denominator's conjugate.

$$\begin{aligned}
\frac{1}{(2 + 3i)(4 - 5i)} &= \frac{1}{23 + 2i} \\
&= \frac{1}{(23 + 2i)} \frac{(23 - 2i)}{(23 - 2i)} \\
&= \frac{23 - 2i}{529 + 4} \\
&= \tfrac{1}{533}(23 - 2i).
\end{aligned}$$

2.7.4 Complex Rotation

Rotate the complex point $3 + 2i$ by $\pm 90°$ and $\pm 180°$.

Solution: Multiply by $\pm i$ and -1.
To rotate $+90°$ (anti-clockwise) multiply by i.

$$i(3 + 2i) = 3i - 2 = -2 + 3i.$$

To rotate $-90°$ (clockwise) multiply by $-i$.

$$-i(3 + 2i) = -3i + 2 = 2 - 3i.$$

To rotate $+180°$ (anti-clockwise) multiply by -1.

$$-1(3 + 2i) = -3 - 2i.$$

To rotate $-180°$ (clockwise) multiply by -1.

$$-1(3 + 2i) = -3 - 2i.$$

2.7.5   Polar Notation

Given $z_1 = \frac{1}{\sqrt{2}} + \frac{\sqrt{2}}{2}i$ and $z_2 = -\frac{1}{\sqrt{2}} + \frac{\sqrt{2}}{2}i$, compute their product using standard complex number format, and polar notation.
Standard complex number format.
Solution: Expand algebraically.

$$z_1 z_2 = \left(\frac{1}{\sqrt{2}} + \frac{\sqrt{2}}{2}i\right)\left(-\frac{1}{\sqrt{2}} + \frac{\sqrt{2}}{2}i\right)$$
$$= -\frac{1}{2} - \frac{1}{2}$$
$$= -1.$$

Polar notation.
Solution: Compute the amplitude and argument for z_1 and z_2; multiply the amplitudes, and add the arguments.

$$r_1 = \sqrt{\left(\frac{1}{\sqrt{2}}\right)^2 + \left(\frac{\sqrt{2}}{2}\right)^2} = 1$$
$$r_2 = \sqrt{\left(-\frac{1}{\sqrt{2}}\right)^2 + \left(\frac{\sqrt{2}}{2}\right)^2} = 1$$
$$\theta_1 = \tan^{-1}\left(\frac{\sqrt{2}}{2}\frac{\sqrt{2}}{1}\right) = 45°$$
$$\theta_2 = \tan^{-1}\left(-\frac{\sqrt{2}}{2}\frac{\sqrt{2}}{1}\right) = 135°$$
$$z_1 = (1, 45°)$$

$$z_2 = (1, 135°)$$
$$z_1 z_2 = (1, 180°) = -1.$$

2.7.6 Real and Imaginary Parts

Find the real and imaginary parts of $1/(1 + e^{i2\theta})$.
Solution: Multiply top and bottom by the conjugate, expand and isolate the real and imaginary parts.

$$\frac{1}{1 + e^{i2\theta}} = \frac{1 + e^{-i2\theta}}{(1 + e^{i2\theta})(1 + e^{-i2\theta})}$$
$$= \frac{1 + \cos(2\theta) - i \sin(2\theta)}{2 + 2\cos(2\theta)}$$
$$= \frac{1 + \cos(2\theta)}{2 + 2\cos(2\theta)} - i \frac{\sin(2\theta)}{2 + 2\cos(2\theta)}$$

$$\text{Re}\left(\frac{1}{1 + e^{i2\theta}}\right) = \frac{1}{2}$$
$$\text{Im}\left(\frac{1}{1 + e^{i2\theta}}\right) = -\frac{1}{2}\left(\frac{\sin(2\theta)}{1 + \cos(2\theta)}\right).$$

2.7.7 Magnitude of a Complex Number

Find $\left|\frac{1}{1+e^{i2\theta}}\right|$.
Solution: Use $z\bar{z} = |z|^2$ and expand.

$$\left|\frac{1}{1 + e^{i2\theta}}\right|^2 = \frac{1}{(1 + e^{i2\theta})(1 + e^{-i2\theta})}$$
$$= \frac{1}{1 + e^{-i2\theta} + e^{i2\theta} + 1}$$
$$= \frac{1}{2 + 2\cos(2\theta)}$$
$$\left|\frac{1}{1 + e^{i2\theta}}\right| = [2 + 2\cos(2\theta)]^{-1/2}.$$

2.7.8   Complex Norm

Find the norm of $z = 5 + 12i$.
Solution: Use $\|z\| = \sqrt{a^2 + b^2}$.

$$\|z\| = |z| = \sqrt{5^2 + 12^2} = 13.$$

2.7.9   Complex Inverse

Find the inverse of $1 + i$.
Solution: Multiply top and bottom by the conjugate and expand.

$$(1 + i)^{-1} = \frac{(1 - i)}{(1 - i)(1 + i)} = \tfrac{1}{2}(1 - i).$$

2.7.10   de Moivre's Theorem

Express $\cos(5\theta)$ in terms of $\cos\theta$, and $\sin(5\theta)$ in terms of $\sin\theta$.
Solution: Use $(\cos\theta + i\sin\theta)^n = \cos(n\theta) + i\sin(n\theta)$ and simplify.

$$\cos(5\theta) + i\sin(5\theta) = (\cos\theta + i\sin\theta)^5$$
$$= \cos^5\theta + 5i\cos^4\theta\sin\theta + 10i^2\cos^3\theta\sin^2\theta$$
$$+ 10i^3\cos^2\theta\sin^3\theta + 5i^4\cos\theta\sin^4\theta + i^5\sin^5\theta$$
$$= \cos^5\theta - 10\cos^3\theta\sin^2\theta + 5\cos\theta\sin^4\theta$$
$$+ i(5\cos^4\theta\sin\theta - 10\cos^2\theta\sin^3\theta + \sin^5\theta)$$
$$\cos(5\theta) = \mathrm{Re}[(\cos\theta + i\sin\theta)^5]$$
$$= \cos^5\theta - 10\cos^3\theta\sin^2\theta + 5\cos\theta\sin^4\theta$$

but $\sin^2\theta = 1 - \cos^2\theta$

$$= \cos^5\theta - 10\cos^3\theta(1 - \cos^2\theta) + 5\cos\theta(1 - \cos^2\theta)^2$$
$$= \cos^5\theta - 10\cos^3\theta + 10\cos^5\theta + 5\cos\theta(1 - 2\cos^2\theta + \cos^4\theta)$$
$$= 11\cos^5\theta - 10\cos^3\theta + 5\cos\theta - 10\cos^3\theta + 5\cos^5\theta$$
$$\cos(5\theta) = 16\cos^5\theta - 20\cos^3\theta + 5\cos\theta.$$

$$\sin(5\theta) = \mathrm{Im}[(\cos\theta + i\sin\theta)^5]$$
$$= 5\cos^4\theta\sin\theta - 10\cos^2\theta\sin^3\theta + \sin^5\theta$$

but $\cos^2\theta = 1 - \sin^2\theta$

$$= 5\sin\theta(1 - \sin^2\theta)^2 - 10\sin^3\theta(1 - \sin^2\theta) + \sin^5\theta$$

Fig. 2.18 $\cos(5\theta) = 16\cos^5\theta - 20\cos^3\theta + 5\cos\theta$

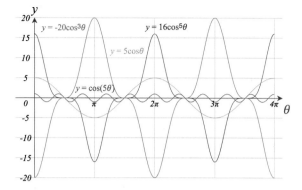

Fig. 2.19 $\sin(5\theta) = 16\sin^5\theta - 20\sin^3\theta + 5\sin\theta$

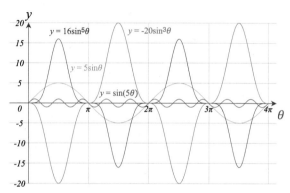

$$= 5\sin\theta - 10\sin^3\theta + 5\sin^5\theta - 10\sin^3\theta + 10\sin^5\theta + \sin^5\theta$$
$$\sin(5\theta) = 16\sin^5\theta - 20\sin^3\theta + 5\sin\theta.$$

Figure 2.18 shows the individual waveforms contributing towards $\cos(5\theta)$, and Fig. 2.19, the individual waveforms contributing towards $\sin(5\theta)$.

2.7.11 nth Root of Unity

Find the 4th and 6th roots of 1.
Solution: Use

$$z = e^{i2k\pi/n}, \quad k = 0, 1, 2, \ldots, n-1$$
$$= \cos\left(\tfrac{2k\pi}{n}\right) + i\sin\left(\tfrac{2k\pi}{n}\right).$$

substituting different values for n and k.

When $n = 4$:

$$[k = 0] \quad z_0 = \cos\left(\tfrac{0}{4}\right) + i \sin\left(\tfrac{0}{4}\right) = 1$$
$$[k = 1] \quad z_1 = \cos\left(\tfrac{2\pi}{4}\right) + i \sin\left(\tfrac{2\pi}{4}\right) = i$$
$$[k = 2] \quad z_1 = \cos\left(\tfrac{4\pi}{4}\right) + i \sin\left(\tfrac{4\pi}{4}\right) = -1$$
$$[k = 3] \quad z_1 = \cos\left(\tfrac{6\pi}{4}\right) + i \sin\left(\tfrac{6\pi}{4}\right) = -i.$$

When $n = 6$:

$$[k = 0] \quad z_0 = \cos\left(\tfrac{0}{6}\right) + i \sin\left(\tfrac{0}{6}\right) = 1$$
$$[k = 1] \quad z_1 = \cos\left(\tfrac{2\pi}{6}\right) + i \sin\left(\tfrac{2\pi}{6}\right) = \tfrac{1}{2} + i\tfrac{\sqrt{3}}{2}$$
$$[k = 2] \quad z_1 = \cos\left(\tfrac{4\pi}{6}\right) + i \sin\left(\tfrac{4\pi}{6}\right) = -\tfrac{1}{2} + i\tfrac{\sqrt{3}}{2}$$
$$[k = 3] \quad z_1 = \cos\left(\tfrac{6\pi}{6}\right) + i \sin\left(\tfrac{6\pi}{6}\right) = -1$$
$$[k = 4] \quad z_1 = \cos\left(\tfrac{8\pi}{6}\right) + i \sin\left(\tfrac{8\pi}{6}\right) = -\tfrac{1}{2} - i\tfrac{\sqrt{3}}{2}$$
$$[k = 5] \quad z_1 = \cos\left(\tfrac{10\pi}{6}\right) + i \sin\left(\tfrac{10\pi}{6}\right) = \tfrac{1}{2} - i\tfrac{\sqrt{3}}{2}.$$

2.7.12   Roots of a Complex Number

Find $\sqrt[3]{i}$.

Solution: Convert i to polar form and use

$$\sqrt[n]{z} = \sqrt[n]{r}\left[\cos\left(\tfrac{\theta + k2\pi}{n}\right) + i \sin\left(\tfrac{\theta + k2\pi}{n}\right)\right], \quad 0 \le k \le n - 1.$$
$$z = 0 + i = (r, \theta)$$
$$r = 1$$
$$\theta = \pi/2$$
$$\sqrt[3]{i} = \cos\left(\tfrac{\theta + k2\pi}{3}\right) + i \sin\left(\tfrac{\theta + k2\pi}{3}\right), \quad 0 \le k \le 2$$
$$= \cos\left(\tfrac{\pi}{6} + \tfrac{k2\pi}{3}\right) + i \sin\left(\tfrac{\pi}{6} + \tfrac{k2\pi}{3}\right)$$
$$z_0 = \cos\left(\tfrac{\pi}{6}\right) + i \sin\left(\tfrac{\pi}{6}\right) = \tfrac{\sqrt{3}}{2} + i\tfrac{1}{2}$$
$$z_1 = \cos\left(\tfrac{\pi}{6} + \tfrac{2\pi}{3}\right) + i \sin\left(\tfrac{\pi}{6} + \tfrac{2\pi}{3}\right) = -\tfrac{\sqrt{3}}{2} + i\tfrac{1}{2}$$
$$z_2 = \cos\left(\tfrac{\pi}{6} + \tfrac{4\pi}{3}\right) + i \sin\left(\tfrac{\pi}{6} + \tfrac{4\pi}{3}\right) = -i.$$

2.7.13 Logarithm of a Complex Number

Compute the natural logarithm of $z = 5 - 12i$.
Solution: Convert z to polar form and use $\ln z = \ln |z| + i \arg(z)$.

$$5 - 12i = \sqrt{5^2 + (-12)^2}\, e^{i \tan^{-1}(-12/5)}$$
$$\approx 13 e^{-i1.176}$$
$$\ln(5 - 12i) \approx \ln\left(13 e^{-i1.176}\right)$$
$$\approx \ln 13 - 1.176i$$
$$\approx 2.565 - 1.176i.$$

2.7.14 Raising a Number to a Complex Power

Compute 3^{1+i}.
Solution: Use $z^w = e^{w \ln z}$.

$$z = 3$$
$$w = 1 + i$$
$$z^w = e^{w \ln z}$$
$$\ln 3 \approx 1.0986$$
$$(1 + i) \ln 3 \approx 1.0986 + 1.0986i$$
$$3^{1+i} \approx e^{(1.0986 + 1.0986i)}$$
$$\approx e^{1.0986} e^{1.0986i}$$
$$\approx 3(\cos 1.0986 + i \sin 1.0986)$$
$$\approx 3(0.4548 + 0.8906i)$$
$$\approx 1.3644 + 2.6718i$$

Reference

1. Feynman RP (1977) The feynman lectures on physics, vol 1. Addison-Wesley, Boston, p 10–22

Chapter 3
Matrix Algebra

3.1 Introduction

This chapter reviews the various types of matrices and their associated algebra
that apply to real and complex numbers. For clarity, they are described in a complex
context. I also describe how a complex number can be represented as a matrix, and
show that $a + bi$, $\cos\theta + i\sin\theta$, $e^{i\theta}$, i and i^{-1}, all have a 2×2 equivalent matrix. If
matrix notation is new to you, then take a look at my book for a complete description
[1].

3.2 Complex Matrices

3.2.1 Matrix Addition and Subtraction

When adding or subtracting matrices, they must be of the same order:

$$C = A \pm B$$
$$\left[c_{ij}\right] = \left[a_{ij}\right] \pm \left[b_{ij}\right]$$

where C has the same order as A and B.

As real and complex number addition is commutative and associative, it follows
that matrix addition is also commutative and associative:

$$A + B = B + A$$
$$(A + B) + C = A + (B + C).$$

For example, given

$$A = \begin{bmatrix} 1 + 2i & 2 + 3i \\ 3 + 4i & 5 - -6i \end{bmatrix}, \quad B = \begin{bmatrix} 1 - 2i & 3 + i \\ 4 - 2i & -2 \end{bmatrix}$$

© Springer International Publishing AG, part of Springer Nature 2018
J. Vince, *Imaginary Mathematics for Computer Science*,
https://doi.org/10.1007/978-3-319-94637-5_3

then

$$A + B = \begin{bmatrix} 1 + 2i & 2 + 3i \\ 3 + 4i & 5 - -6i \end{bmatrix} + \begin{bmatrix} 1 - 2i & 3 + i \\ 4 - 2i & -2 \end{bmatrix}$$

$$= \begin{bmatrix} 2 & 5 + 4i \\ 7 + 2i & 3 - 6i \end{bmatrix}$$

$$A - B = \begin{bmatrix} 1 + 2i & 2 + 3i \\ 3 + 4i & 5 - 6i \end{bmatrix} - \begin{bmatrix} 1 - 2i & 3 + i \\ 4 - 2i & -2 \end{bmatrix}$$

$$= \begin{bmatrix} 4i & -1 + 2i \\ -1 + 6i & 7 - 6i \end{bmatrix}.$$

3.2.2   Matrix Scaling

Matrix scaling is the action of multiplying each element of a matrix by a scaling factor, real or complex. For example, matrix A is scaled by λ as follows:

$$\lambda A = \lambda \begin{bmatrix} a_{ij} \end{bmatrix} = \begin{bmatrix} \lambda a_{ij} \end{bmatrix}$$

where each element of A is multiplied by λ. If $\lambda = 2i$ then

$$A = \begin{bmatrix} 1 + i & 2 - 3i \\ 3 - 2i & 4 \end{bmatrix}$$

$$\lambda A = 2i \begin{bmatrix} 1 + i & 2 - 3i \\ 3 - 2i & 4 \end{bmatrix} = \begin{bmatrix} -2 + 2i & 6 + 4i \\ 4 + 6i & 8i \end{bmatrix}.$$

It follows that if the elements of a matrix share a common factor, the factor can be placed outside the matrix. For example,

$$B = \begin{bmatrix} 10i & 20i \\ 30i & 40i \end{bmatrix}$$

$$= i \begin{bmatrix} 10 & 20 \\ 30 & 40 \end{bmatrix}.$$

3.2.3   Zero Matrix

By definition, all the elements of a *zero matrix* equal zero and is represented by **0**. Here are some examples,

$$[0 \ 0], \quad [0 \ 0 \ 0], \quad \begin{bmatrix} 0 \\ 0 \end{bmatrix}, \quad \begin{bmatrix} 0 \\ 0 \\ 0 \end{bmatrix}, \quad \begin{bmatrix} 0 & 0 & 0 \\ 0 & 0 & 0 \\ 0 & 0 & 0 \end{bmatrix}.$$

It follows from the rules of matrix addition that $A + \mathbf{0} = A$.

3.2.4 Matrix Multiplication

For matrix multiplication to be consistent with its algebraic equivalent, matrix multiplication must obey certain rules. For instance, given two matrices A and B,

$$A = \begin{bmatrix} a_{11} & a_{12} \\ a_{21} & a_{22} \end{bmatrix}, \quad B = \begin{bmatrix} b_{11} & b_{12} \\ b_{21} & b_{22} \end{bmatrix}$$

then

$$AB = \begin{bmatrix} a_{11}b_{11} + a_{12}b_{21} & a_{11}b_{12} + a_{12}b_{22} \\ a_{21}b_{11} + a_{22}b_{21} & a_{21}b_{12} + a_{22}b_{22} \end{bmatrix}.$$

This can be generalised as follows. Given two matrices A and B, where A is a matrix of order $m \times p$ with elements a_{ik}, and B is a matrix of order $p \times n$ with elements b_{kj}, then $C = AB$ is a matrix of order $m \times n$ with elements c_{ij}, where

$$c_{ij} = a_{i1}b_{1j} + a_{i2}b_{2j} + a_{i3}b_{3j} + \cdots + a_{ip}b_{pj},$$

which can be expressed as

$$c_{ij} = \sum_{k=1}^{p} a_{ik}b_{kj}. \tag{3.1}$$

For example, two complex matrices of order 2 are multiplied as follows:

$$A = \begin{bmatrix} 1+i & 2-2i \\ 4 & 5+4i \end{bmatrix}, \quad B = \begin{bmatrix} 3+i & 4-i \\ 2+3i & 5+2i \end{bmatrix}.$$

$$\begin{aligned} C &= AB \\ &= \begin{bmatrix} 1+i & 2-2i \\ 4 & 5+4i \end{bmatrix} \begin{bmatrix} 3+i & 4-i \\ 2+3i & 5+2i \end{bmatrix} \\ &= \begin{bmatrix} (1+i)(3+i) + (2-2i)(2+3i) & (1+i)(4-i) + (2-2i)(5+2i) \\ 4(3+i) + (5+4i)(2+3i) & 4(4-i) + (5+4i)(5+2i) \end{bmatrix} \end{aligned}$$

$$= \begin{bmatrix} (2+4i) + (10+2i) & (5+3i) + (14-6i) \\ (12+4i) + (-2+23i) & (16-4i) + (17+30i) \end{bmatrix}$$

$$= \begin{bmatrix} 12+6i & 19-3i \\ 10+27i & 33+26i \end{bmatrix}.$$

Reversing the product such that $C = BA$, changes every element of C and is the reason why, in general, matrix multiplication is non-commutative.

3.2.5 Negative Matrix

By definition, given a matrix A with elements a_{ij}, its negative, $-A$ is defined such that

$$-A = [-a_{ij}].$$

For example, given

$$A = \begin{bmatrix} 1+i & -2-3i & 3i \\ -4 & 5+2i & -6+4i \\ 7i & -8 & 9-3i \end{bmatrix}$$

then

$$-A = \begin{bmatrix} -1-i & 2+3i & -3i \\ 4 & -5-2i & 6-4i \\ -7i & 8 & -9+3i \end{bmatrix}.$$

It follows that $A + (-A) = \mathbf{0}$, because

$$A + (-A) = [a_{ij}] + [-a_{ij}] = [0_{ij}].$$

3.2.6 Determinant of a Matrix

When a 2×2 matrix is used as a geometric transform, its determinant represents the scaling action. For instance, the determinant of the identity matrix is 1:

$$\det \begin{bmatrix} 1 & 0 \\ 0 & 1 \end{bmatrix} = \begin{vmatrix} 1 & 0 \\ 0 & 1 \end{vmatrix} = 1.$$

The determinant of a rotation matrix should also equal 1, as no scaling takes place,

$$\det \begin{bmatrix} \cos\theta & -\sin\theta \\ \sin\theta & \cos\theta \end{bmatrix} = \begin{vmatrix} \cos\theta & -\sin\theta \\ \sin\theta & \cos\theta \end{vmatrix} = \cos^2\theta + \sin^2\theta = 1.$$

However, given a complex matrix, its determinant is probably complex, but not necessarily so. Let's start with the following matrix

$$A = \begin{bmatrix} 2+3i & 3+2i \\ 1-i & 4-2i \end{bmatrix}$$

then

$$\begin{aligned} |A| &= (2+3i)(4-2i) - (1-i)(3+2i) \\ &= (14+8i) - (5-i) \\ &= 9+9i. \end{aligned}$$

whereas, given

$$B = \begin{bmatrix} 4+3i & 3+2i \\ 3-2i & 4-3i \end{bmatrix}$$

then

$$\begin{aligned} |B| &= (4+3i)(4-3i) - (3-2i)(3+2i) \\ &= 23 - 13 \\ &= 10. \end{aligned}$$

3.2.7 Diagonal Matrix

A *diagonal matrix* is an $n \times n$ matrix whose elements are zero, apart from its diagonal:

$$A = \begin{bmatrix} a_{11} & 0 & \cdots & 0 \\ 0 & a_{22} & \cdots & 0 \\ \vdots & \vdots & \ddots & \vdots \\ 0 & 0 & \cdots & a_{nn} \end{bmatrix},$$

consequently, the determinant of a diagonal matrix must be

$$|A| = a_{11} \times a_{22} \times \cdots \times a_{nn}.$$

Here is a diagonal matrix with its determinant

$$A = \begin{bmatrix} 2+3i & 0 & 0 \\ 0 & 3-i & 0 \\ 0 & 0 & 4+5i \end{bmatrix}$$

$$\begin{aligned} |A| &= (2+3i)(3-i)(4+5i) \\ &= (2+3i)(17+11i) \\ &= 1+73i. \end{aligned}$$

Now let's consider the product of two diagonal matrices A and B with the same order. The general product rule is

$$c_{ij} = a_{i1}b_{1j} + a_{i2}b_{2j} + a_{i3}b_{3j} + \cdots + a_{in}b_{nj},$$

for all ij pairs. As the rows of A multiply the columns of B, the only time there will be a non-zero result, is when the row and column share a common diagonal element. Consequently, the resulting product is also a diagonal matrix. Let's illustrate this with a 3×3 matrix:

$$A = \begin{bmatrix} a_{11} & 0 & 0 \\ 0 & a_{22} & 0 \\ 0 & 0 & a_{33} \end{bmatrix}, \quad B = \begin{bmatrix} b_{11} & 0 & 0 \\ 0 & b_{22} & 0 \\ 0 & 0 & b_{33} \end{bmatrix}$$

$$AB = \begin{bmatrix} a_{11} & 0 & 0 \\ 0 & a_{22} & 0 \\ 0 & 0 & a_{33} \end{bmatrix} \begin{bmatrix} b_{11} & 0 & 0 \\ 0 & b_{22} & 0 \\ 0 & 0 & b_{33} \end{bmatrix}$$

$$= \begin{bmatrix} a_{11}b_{11} & 0 & 0 \\ 0 & a_{22}b_{22} & 0 \\ 0 & 0 & a_{33}b_{33} \end{bmatrix}$$

which is another diagonal matrix.

Here is an example:

$$A = \begin{bmatrix} 2+i & 0 & 0 \\ 0 & 4 & 0 \\ 0 & 0 & 6i \end{bmatrix}, \quad B = \begin{bmatrix} 3 & 0 & 0 \\ 0 & 5+i & 0 \\ 0 & 0 & 7-i \end{bmatrix}$$

$$AB = \begin{bmatrix} 6+3i & 0 & 0 \\ 0 & 20+4i & 0 \\ 0 & 0 & 6+42i \end{bmatrix}.$$

It should be clear that matrix multiplication of diagonal matrices is commutative. i.e. $AB = BA$.

3.2.8 Identity Matrix

An *identity matrix* is a diagonal matrix with its elements equal to 1. For example:

$$I_2 = \begin{bmatrix} 1 & 0 \\ 0 & 1 \end{bmatrix}, \quad I_3 = \begin{bmatrix} 1 & 0 & 0 \\ 0 & 1 & 0 \\ 0 & 0 & 1 \end{bmatrix}, \quad I_4 = \begin{bmatrix} 1 & 0 & 0 & 0 \\ 0 & 1 & 0 & 0 \\ 0 & 0 & 1 & 0 \\ 0 & 0 & 0 & 1 \end{bmatrix}.$$

Thus when we multiply any matrix by I_n it leaves the matrix unchanged. It is common to employ I as the identity matrix, as its order is determined by the associated matrix. Note that $IA = AI$.

3.2.9 Transpose Matrix

A *transpose matrix* is denoted by A^T and exchanges rows with columns in the original matrix. Only the diagonal elements remain unchanged. For example:

$$A = \begin{bmatrix} 1+i & 2+2i & 3+3i \\ 4+4i & 5+5i & 6+6i \\ 7+7i & 8+8i & 9+9i \end{bmatrix}$$

then

$$A^T = \begin{bmatrix} 1+i & 4+4i & 7+7i \\ 2+2i & 5+5i & 8+8i \\ 3+3i & 6+6i & 9+9i \end{bmatrix}.$$

3.2.10 Trace

The *trace* of a square matrix A is defined as the sum of its diagonal elements and written as $\text{Tr}(A)$. For example, given

$$A = \begin{bmatrix} 1+i & 4+4i & 7+7i \\ 2+2i & 5+5i & 8+8i \\ 3+3i & 6+6i & 9+9i \end{bmatrix}$$

$$\text{Tr}(A) = (1+i) + (5+5i) + (9+9i) = 15 + 15i.$$

3.2.11   Symmetric Matrix

It is worth exploring two types of matrices called *symmetric* and *anti-symmetric* matrices, as we refer to them later on. A symmetric matrix is a matrix which equals its own transpose:

$$A = A^{\mathrm{T}}.$$

For example, the following matrix is symmetric:

$$A = \begin{bmatrix} 1+2i & 3+4i & 4+5i \\ 3+4i & 2+3i & 6+7i \\ 4+5i & 6+7i & 3+4i \end{bmatrix}.$$

The symmetric part of any square matrix can be isolated as follows. Given a matrix A and its transpose A^{T}:

$$A = \begin{bmatrix} a_{11} & a_{12} & \cdots & a_{1n} \\ a_{21} & a_{22} & \cdots & a_{2n} \\ \vdots & \vdots & \ddots & \vdots \\ a_{n1} & a_{n2} & \cdots & a_{nn} \end{bmatrix}, \quad A^{\mathrm{T}} = \begin{bmatrix} a_{11} & a_{21} & \cdots & a_{n1} \\ a_{12} & a_{22} & \cdots & a_{n2} \\ \vdots & \vdots & \ddots & \vdots \\ a_{1n} & a_{2n} & \cdots & a_{nn} \end{bmatrix}$$

their sum is

$$A + A^{\mathrm{T}} = \begin{bmatrix} 2a_{11} & a_{12}+a_{21} & \cdots & a_{1n}+a_{n1} \\ a_{12}+a_{21} & 2a_{22} & \cdots & a_{2n}+a_{n2} \\ \vdots & \vdots & \ddots & \vdots \\ a_{1n}+a_{n1} & a_{2n}+a_{n2} & \cdots & 2a_{nn} \end{bmatrix}. \tag{3.2}$$

By inspection, (3.2) is symmetric, and dividing by 2 we have

$$S = \tfrac{1}{2}\left(A + A^{\mathrm{T}}\right),$$

which is defined as the symmetric part of A.

For example, given

$$A = \begin{bmatrix} a_{11} & a_{12} & a_{13} \\ a_{21} & a_{22} & a_{23} \\ a_{31} & a_{32} & a_{33} \end{bmatrix}, \quad A^{\mathrm{T}} = \begin{bmatrix} a_{11} & a_{21} & a_{31} \\ a_{12} & a_{22} & a_{32} \\ a_{13} & a_{23} & a_{33} \end{bmatrix}$$

then

$$S = \tfrac{1}{2}\left(A + A^{\mathrm{T}}\right)$$

$$= \begin{bmatrix} a_{11} & \dfrac{a_{12} + a_{21}}{2} & \dfrac{a_{13} + a_{31}}{2} \\ \dfrac{a_{12} + a_{21}}{2} & a_{22} & \dfrac{a_{23} + a_{32}}{2} \\ \dfrac{a_{13} + a_{31}}{2} & \dfrac{a_{23} + a_{32}}{2} & a_{33} \end{bmatrix}$$

$$= \begin{bmatrix} a_{11} & \dfrac{s_3}{2} & \dfrac{s_2}{2} \\ \dfrac{s_3}{2} & a_{22} & \dfrac{s_1}{2} \\ \dfrac{s_2}{2} & \dfrac{s_1}{2} & a_{33} \end{bmatrix}$$

where

$$s_1 = a_{23} + a_{32}$$
$$s_2 = a_{13} + a_{31}$$
$$s_3 = a_{12} + a_{21}.$$

For example, given

$$A = \begin{bmatrix} 0 & 1+i & 4+4i \\ 3-i & 2+i & -3i \\ 4+4i & 2+3i & 6+2i \end{bmatrix}, \quad A^{\mathrm{T}} = \begin{bmatrix} 0 & 3-i & 4+4i \\ 1+i & 2+i & 2+3i \\ 4+4i & -3i & 6+2i \end{bmatrix}$$

$$s_1 = 2, \quad s_2 = 8 + 8i, \quad s_3 = 4$$

$$S = \begin{bmatrix} 0 & 2 & 4+4i \\ 2 & 2+i & 1 \\ 4+4i & 1 & 6+2i \end{bmatrix}$$

which equals its own transpose.

3.2.12 Anti-symmetric Matrix

An *anti-symmetric matrix* is a matrix whose transpose is its own negative:

$$A^{\mathrm{T}} = -A,$$

and is also known as a *skew symmetric matrix*.

As the elements of A and A^{T} are related by

$$a_{ij} = -a_{ji},$$

when $k = i = j$

$$a_{kk} = -a_{kk},$$

which implies that the diagonal elements must be zero. For example, this is an anti-symmetric matrix

$$\begin{bmatrix} 0 & 6+5i & 2+3i \\ -6-5i & 0 & 4+5i \\ -2-3i & -4-5i & 0 \end{bmatrix}.$$

In general, we have

$$A = \begin{bmatrix} a_{11} & a_{12} & \cdots & a_{1n} \\ a_{21} & a_{22} & \cdots & a_{2n} \\ \vdots & \vdots & \ddots & \vdots \\ a_{n1} & a_{n2} & \cdots & a_{nn} \end{bmatrix}, \quad A^{\mathrm{T}} = \begin{bmatrix} a_{11} & a_{21} & \cdots & a_{n1} \\ a_{12} & a_{22} & \cdots & a_{n2} \\ \vdots & \vdots & \ddots & \vdots \\ a_{1n} & a_{2n} & \cdots & a_{nn} \end{bmatrix}$$

and their difference is

$$A - A^{\mathrm{T}} = \begin{bmatrix} 0 & a_{12} - a_{21} & \cdots & a_{1n} - a_{n1} \\ -(a_{12} - a_{21}) & 0 & \cdots & a_{2n} - a_{n2} \\ \vdots & \vdots & \ddots & \vdots \\ -(a_{1n} - a_{n1}) & -(a_{2n} - a_{n2}) & \cdots & 0 \end{bmatrix}. \tag{3.3}$$

It is clear that (3.3) is anti-symmetric, and dividing by 2 we have

$$Q = \tfrac{1}{2}\left(A - A^{\mathrm{T}}\right).$$

For example,

$$A = \begin{bmatrix} a_{11} & a_{12} & a_{13} \\ a_{21} & a_{22} & a_{23} \\ a_{31} & a_{32} & a_{33} \end{bmatrix}, \quad A^{\mathrm{T}} = \begin{bmatrix} a_{11} & a_{21} & a_{31} \\ a_{12} & a_{22} & a_{32} \\ a_{13} & a_{23} & a_{33} \end{bmatrix}$$

$$Q = \begin{bmatrix} 0 & \dfrac{a_{12} - a_{21}}{2} & \dfrac{a_{13} - a_{31}}{2} \\ \dfrac{a_{21} - a_{12}}{2} & 0 & \dfrac{a_{23} - a_{32}}{2} \\ \dfrac{a_{31} - a_{13}}{2} & \dfrac{a_{32} - a_{23}}{2} & 0 \end{bmatrix}$$

and if we maintain some symmetry with the subscripts, we have

$$
Q = \begin{bmatrix} 0 & \dfrac{a_{12} - a_{21}}{2} & -\dfrac{a_{31} - a_{13}}{2} \\ -\dfrac{a_{12} - a_{21}}{2} & 0 & \dfrac{a_{23} - a_{32}}{2} \\ \dfrac{a_{31} - a_{13}}{2} & -\dfrac{a_{23} - a_{32}}{2} & 0 \end{bmatrix}
$$

$$
= \begin{bmatrix} 0 & \dfrac{q_3}{2} & -\dfrac{q_2}{2} \\ -\dfrac{q_3}{2} & 0 & \dfrac{q_1}{2} \\ \dfrac{q_2}{2} & -\dfrac{q_1}{2} & 0 \end{bmatrix}
$$

where

$$q_1 = a_{23} - a_{32}$$
$$q_2 = a_{31} - a_{13}$$
$$q_3 = a_{12} - a_{21}.$$

For example,

$$
A = \begin{bmatrix} 0 & 1+i & 4+4i \\ 3-i & 2+i & -3i \\ 4+4i & 2+3i & 6+2i \end{bmatrix}, \quad A^{\mathrm{T}} = \begin{bmatrix} 0 & 3-i & 4+4i \\ 1+i & 2+i & 2+3i \\ 4+4i & -3i & 6+2i \end{bmatrix}
$$

$$q_1 = -2 - 6i, \quad q_2 = 0, \quad q_3 = -2 + 2i$$

$$
Q = \begin{bmatrix} 0 & -1+i & 0 \\ 1-i & 0 & -1-3i \\ 0 & 1+3i & 0 \end{bmatrix}.
$$

Furthermore, we have already computed

$$
S = \begin{bmatrix} 0 & 2 & 4+4i \\ 2 & 2+i & 1 \\ 4+4i & 1 & 6+2i \end{bmatrix}
$$

and

$$
S + Q = \begin{bmatrix} 0 & 1+i & 4+4i \\ 3-i & 2+i & -3i \\ 4+4i & 2+3i & 6+2i \end{bmatrix} = A.
$$

3.2.13 Inverse Matrix

One of the useful features of matrix notation is the concept of the *inverse matrix* where a square matrix A, may have an inverse A^{-1}, such that the product $AA^{-1} = I$. In the world of transforms, a matrix A performs a transformation such as a rotation about an axis, whilst its inverse A^{-1}, performs the inverse transformation, which rotates in the opposite direction.

So a useful definition for an inverse matrix is: Let A be a square matrix of order n, and A^{-1} be another square matrix of order n, such that their product $AA^{-1} = A^{-1}A = I$. This definition preempts the possibility of matrices that do not have an inverse. For example, the following matrix does not have an inverse, as its determinant is zero:

$$A = \begin{bmatrix} 1+i & 1+i \\ 1+i & 1+i \end{bmatrix}$$
$$|A| = (1+i)(1+i) - (1+i)(1+i)$$
$$= 0.$$

Therefore, from now on, when we talk about an inverse matrix, we assume the existence of the inverse form.

One way to derive an inverse matrix employs a cofactor matrix, which is based upon the cofactors associated with any matrix element.

3.2.14 Cofactor Matrix

Let's start with the following matrix and its cofactor matrix

$$A = \begin{bmatrix} i & 1+i & 3-3i \\ 2+2i & 1-i & 4+4i \\ 4-4i & 2+i & 2-2i \end{bmatrix}$$

$$\text{cofactor matrix of } A = \begin{bmatrix} A_{11} & A_{12} & A_{13} \\ A_{21} & A_{22} & A_{23} \\ A_{31} & A_{32} & A_{33} \end{bmatrix}$$

where

$$A_{11} = + \begin{vmatrix} a_{22} & a_{23} \\ a_{32} & a_{33} \end{vmatrix} = + \begin{vmatrix} 1-i & 4+4i \\ 2+i & 2-2i \end{vmatrix} = -4 - 16i$$

$$A_{12} = - \begin{vmatrix} a_{21} & a_{23} \\ a_{31} & a_{33} \end{vmatrix} = - \begin{vmatrix} 2+2i & 4+4i \\ 4-4i & 2-2i \end{vmatrix} = 24$$

$$A_{13} = + \begin{vmatrix} a_{21} & a_{22} \\ a_{31} & a_{32} \end{vmatrix} = + \begin{vmatrix} 2 + 2i & 1 - i \\ 4 - 4i & 2 + i \end{vmatrix} = 2 + 14i$$

$$A_{21} = - \begin{vmatrix} a_{12} & a_{13} \\ a_{32} & a_{33} \end{vmatrix} = - \begin{vmatrix} 1 + i & 3 - 3i \\ 2 + i & 2 - 2i \end{vmatrix} = 5 - 3i$$

$$A_{22} = + \begin{vmatrix} a_{11} & a_{13} \\ a_{31} & a_{33} \end{vmatrix} = + \begin{vmatrix} i & 3 - 3i \\ 4 - 4i & 2 - 2i \end{vmatrix} = 2 + 26i$$

$$A_{23} = - \begin{vmatrix} a_{11} & a_{12} \\ a_{31} & a_{32} \end{vmatrix} = - \begin{vmatrix} i & 1 + i \\ 4 - 4i & 2 + i \end{vmatrix} = 9 - 2i$$

$$A_{31} = + \begin{vmatrix} a_{12} & a_{13} \\ a_{22} & a_{23} \end{vmatrix} = + \begin{vmatrix} 1 + i & 3 - 3i \\ 1 - i & 4 + 4i \end{vmatrix} = 14i$$

$$A_{32} = - \begin{vmatrix} a_{11} & a_{13} \\ a_{21} & a_{23} \end{vmatrix} = - \begin{vmatrix} i & 3 - 3i \\ 2 + 2i & 4 + 4i \end{vmatrix} = 16 - 4i$$

$$A_{33} = + \begin{vmatrix} a_{11} & a_{12} \\ a_{21} & a_{22} \end{vmatrix} = + \begin{vmatrix} i & 1 + i \\ 2 + 2i & 1 - i \end{vmatrix} = 1 - 3i$$

therefore, the cofactor matrix of A is

$$\begin{bmatrix} -4 - 16i & 24 & 2 + 14i \\ 5 - 3i & 2 + 26i & 9 - 2i \\ 14i & 16 - 4i & 1 - 3i \end{bmatrix}.$$

It can be shown that the product of a matrix with the transpose of its cofactor matrix has the following form:

$$A(\text{cofactor matrix of } A)^{\mathrm{T}} = \begin{bmatrix} |A| & 0 & \cdots & 0 \\ 0 & |A| & \cdots & 0 \\ \vdots & \vdots & \ddots & \vdots \\ 0 & 0 & 0 & |A| \end{bmatrix}.$$

and dividing throughout by $|A|$ we have

$$\frac{A(\text{cofactor matrix of } A)^{\mathrm{T}}}{|A|} = I,$$

which implies that

$$A^{-1} = \frac{(\text{cofactor matrix of } A)^{\mathrm{T}}}{|A|}.$$

Naturally, this assumes that the inverse actually exists.

Let's find the inverse of the above matrix

$$A = \begin{bmatrix} i & 1+i & 3-3i \\ 2+2i & 1-i & 4+4i \\ 4-4i & 2+i & 2-2i \end{bmatrix}$$

$$(\text{cofactor matrix of } A) = \begin{bmatrix} -4-16i & 24 & 2+14i \\ 5-3i & 2+26i & 9-2i \\ 14i & 16-4i & 1-3i \end{bmatrix}$$

$$(\text{cofactor matrix of } A)^{\mathrm{T}} = \begin{bmatrix} -4-16i & 5-3i & 14i \\ 24 & 2+26i & 16-4i \\ 2+14i & 9-2i & 1-3i \end{bmatrix}$$

$$|A| = a_{11}A_{11} + a_{12}A_{12} + a_{13}A_{13}$$
$$= i(-4-16i) + (1+i)(24) + (3-3i)(2+14i)$$
$$= 16 - 4i + 24 + 24i + 48 + 36i$$
$$= 88 + 56i$$

$$A^{-1} = \frac{1}{88+56i} \begin{bmatrix} -4-16i & 5-3i & 14i \\ 24 & 2+26i & 16-4i \\ 2+14i & 9-2i & 1-3i \end{bmatrix}.$$

Let's check this result by multiplying A by A^{-1} which must equal I:

$$AA^{-1} = \frac{1}{88+56i} \begin{bmatrix} i & 1+i & 3-3i \\ 2+2i & 1-i & 4+4i \\ 4-4i & 2+i & 2-2i \end{bmatrix} \begin{bmatrix} -4-16i & 5-3i & 14i \\ 24 & 2+26i & 16-4i \\ 2+14i & 9-2i & 1-3i \end{bmatrix}$$

$$= \frac{1}{88+56i} \begin{bmatrix} 88+56i & 0 & 0 \\ 0 & 88+56i & 0 \\ 0 & 0 & 88+56i \end{bmatrix}$$

$$= \begin{bmatrix} 1 & 0 & 0 \\ 0 & 1 & 0 \\ 0 & 0 & 1 \end{bmatrix}.$$

Finally, let's compute the inverse matrix of the following matrix using cofactors:

$$A = \begin{bmatrix} 2+i & 3-2i \\ 4-i & -1+i \end{bmatrix}$$

$$(\text{cofactor matrix of } A) = \begin{bmatrix} -1+i & -4+i \\ -3+2i & 2+i \end{bmatrix}$$

$$(\text{cofactor matrix of } A)^{\mathrm{T}} = \begin{bmatrix} -1+i & -3+2i \\ -4+i & 2+i \end{bmatrix}$$

$$|A| = (2+i)(-1+i) + (3-2i)(-4+i) = -13+12i$$

$$A^{-1} = \frac{1}{-13+12i} \begin{bmatrix} -1+i & -3+2i \\ -4+i & 2+i \end{bmatrix}.$$

Let's test it:

$$A^{-1}A = \frac{1}{-13 + 12i} \begin{bmatrix} -1+i & -3+2i \\ -4+i & 2+i \end{bmatrix} \begin{bmatrix} 2+i & 3-2i \\ 4-i & -1+i \end{bmatrix}$$

$$= \frac{1}{-13 + 12i} \begin{bmatrix} -13 + 12i & 0 \\ 0 & -13 + 12i \end{bmatrix}$$

$$= \begin{bmatrix} 1 & 0 \\ 0 & 1 \end{bmatrix}.$$

In general, the inverse of a 2×2 matrix is given by

$$A = \begin{bmatrix} a_{11} & a_{12} \\ a_{21} & a_{22} \end{bmatrix}$$

$$A^{-1} = \frac{1}{a_{11}a_{22} - a_{12}a_{21}} \begin{bmatrix} a_{22} & -a_{12} \\ -a_{21} & a_{11} \end{bmatrix}.$$

3.2.15 Conjugate Matrix

By definition, the conjugate of a complex number $z = a + bi$ is $\bar{z} = a - bi$. Similarly, given an $n \times m$ matrix containing complex entries, its conjugate is an $n \times m$ matrix with the sign of the imaginary parts reversed. For example, given A

$$A = \begin{bmatrix} 2+3i & 4-5i \\ -i & 1+2i \end{bmatrix}$$

its conjugate, $\bar{A}$ is

$$\bar{A} = \begin{bmatrix} 2-3i & 4+5i \\ i & 1-2i \end{bmatrix}.$$

If $A = \bar{A}$ then A must be a real matrix.

3.2.16 Normal Matrix

A complex square matrix is *normal* when

$$\overline{A^T}A = A\overline{A^T}$$

whereas a real square matrix is normal when

$$A^T A = A A^T.$$

For example, A is normal:

$$A = \begin{bmatrix} 1 & 4i & 5 \\ -4i & 2 & 3+3i \\ 5 & 3-3i & 3 \end{bmatrix}$$

$$A^T = \begin{bmatrix} 1 & -4i & 5 \\ 4i & 2 & 3-3i \\ 5 & 3+3i & 3 \end{bmatrix}$$

$$\overline{A^T} = \begin{bmatrix} 1 & 4i & 5 \\ -4i & 2 & 3+3i \\ 5 & 3-3i & 3 \end{bmatrix}$$

$$\overline{A^T} A = \begin{bmatrix} 1 & 4i & 5 \\ -4i & 2 & 3+3i \\ 5 & 3-3i & 3 \end{bmatrix} \begin{bmatrix} 1 & 4i & 5 \\ -4i & 2 & 3+3i \\ 5 & 3-3i & 3 \end{bmatrix} = A\overline{A^T}.$$

3.2.17   Conjugate Transpose

Given an $n \times m$ matrix A, its *conjugate transpose* or *Hermitian transpose* A^* is achieved by transposing A and then, conjugating all the entries:

$$A^* = \overline{A^T}.$$

In fact, it should be obvious that reversing the transposition and conjugation has no effect on the result:

$$A^* = \overline{A}^T.$$

For example, given

$$A = \begin{bmatrix} 1 & 4i & -2i \\ -4i & 2 & 3+3i \end{bmatrix}$$

then

$$\overline{A} = \begin{bmatrix} 1 & -4i & 2i \\ 4i & 2 & 3-3i \end{bmatrix}$$

and

$$A^* = \begin{bmatrix} 1 & 4i \\ -4i & 2 \\ 2i & 3-3i \end{bmatrix}.$$

Here are some observations regarding such matrices:

1. Reversing the conjugate and transpose operations has no effect on the result.
2. $(A + B)^* = A^* + B^*$, where A and B are $n \times n$ matrices.
3. $(AB)^* = B^*A^*$, for any $m \times n$ matrix A and any $n \times p$ matrix B.
4. $(A^*)^* = A$, for any $m \times n$ matrix A.
5. $(zA)^* = \bar{z}A^*$, for any complex number z and any $m \times n$ matrix A.
6. Other names for the conjugate transpose are *Hermitian conjugate* or *adjoint matrix*.

3.2.18 Hermitian Matrix

A *Hermitian matrix* is a complex square matrix that is equal to its own conjugate transpose, and is denoted A^{H}. i.e. $A = \overline{A}^{\mathrm{T}}$, or $A = \overline{A^{\mathrm{T}}}$. It is named after the French mathematician Charles Hermite (1822–1901), who made a significant contribution to mathematics, including a proof for e being a transcendental number. Let's illustrate this with an example. Given,

$$A = \begin{bmatrix} 1 & 4i & 5 \\ -4i & 2 & 3+3i \\ 5 & 3-3i & 3 \end{bmatrix}$$

then

$$\overline{A} = \begin{bmatrix} 1 & -4i & 5 \\ 4i & 2 & 3-3i \\ 5 & 3+3i & 3 \end{bmatrix}$$

whose transpose $\overline{A}^{\mathrm{T}}$ gives the original matrix, A:

$$A = \begin{bmatrix} 1 & 4i & 5 \\ -4i & 2 & 3+3i \\ 5 & 3-3i & 3 \end{bmatrix}$$

and makes it a Hermitian matrix.

An important property with Hermitian matrices is that they always have real eigenvalues. For example, matrix A is Hermitian, and has real eigenvalues $\lambda = 1.5 \pm \sqrt{65}/2$:

$$A = \begin{bmatrix} 1 & 4i \\ -4i & 2 \end{bmatrix}$$

and solving the characteristic equation:

$$\begin{vmatrix} 1 - \lambda & 4i \\ -4i & 2 - \lambda \end{vmatrix} = 0$$

$$(1 - \lambda)(2 - \lambda) - 16 = 0$$

$$\lambda^2 - 3\lambda - 14 = 0$$

$$\lambda = \frac{3 \pm \sqrt{9 + 56}}{2}$$

$$\lambda = 1.5 \pm \sqrt{65}/2.$$

Let's also show that the above 3×3 Hermitian matrix has real eigenvalues.

$$A = \begin{bmatrix} 1 & 4i & 5 \\ -4i & 2 & 3 + 3i \\ 5 & 3 - 3i & 3 \end{bmatrix}$$

solving the characteristic equation:

$$\begin{vmatrix} 1 - \lambda & 4i & 5 \\ -4i & 2 - \lambda & 3 + 3i \\ 5 & 3 - 3i & 3 - \lambda \end{vmatrix} = 0.$$

Using Sarrus's rule:

$$\begin{aligned} 0 &= (1 - \lambda)(2 - \lambda)(3 - \lambda) + 4i(3 + 3i)5 + 5(-4i)(3 - 3i) - \\ &\quad (1 - \lambda)(3 + 3i)(3 - 3i) - 4i(-4i)(3 - \lambda) - 5(2 - \lambda)5 \\ &= (2 - 3\lambda + \lambda^2)(3 - \lambda) - 60 + 60i - 60 - 60i - \\ &\quad (1 - \lambda)18 - (48 - 16\lambda) - (50 - 25\lambda) \\ &= -\lambda^3 + 6\lambda^2 - 11\lambda + 6 - 120 - \\ &\quad 18 + 18\lambda - 48 + 16\lambda - 50 + 25\lambda \\ &= -\lambda^3 + 6\lambda^2 + 48\lambda - 230 \\ &= \lambda^3 - 6\lambda^2 - 48\lambda + 230. \end{aligned}$$

Fig. 3.1 Graph of
$y = \lambda^3 - 6\lambda^2 - 48\lambda + 230$

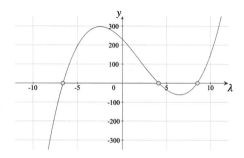

Figure 3.1 shows the graph of $y = \lambda^3 - 6\lambda^2 - 48\lambda + 230$, with real solutions: $\lambda \approx$ $-6.6,\ 4.1,\ 8.5$.

Hermitian matrices are rather special and possess many properties:

1. The main diagonal entries have to be real, because they must equal their complex conjugate.
2. Charles Hermite proved in 1855 that matrices, where $A = \overline{A}^{\mathrm{T}}$, always have real eigenvalues. He didn't use this term, as this was coined by Hilbert, 50 years later.
3. Hermitian matrices are normal matrices.
4. $A_1^{\mathrm{H}} + A_2^{\mathrm{H}} = A_3^{\mathrm{H}}$.
5. If A^{H} has an inverse, it is also Hermitian.
6. A matrix such that $A^{\mathrm{H}} = -A$ is a *skew-Hermitian matrix*. See www.wikipedia. org for a complete overview.

3.2.19   Orthogonal Matrix

The transpose of an *orthogonal matrix* is also its inverse:

$$A^{\mathrm{T}} = A^{-1}$$

which makes inverting extremely easy. For example, the following 2D rotation transform is orthogonal:

$$A = \begin{bmatrix} \cos\theta & -\sin\theta \\ \sin\theta & \cos\theta \end{bmatrix}$$

because its transpose is its inverse:

$$A^{\mathrm{T}} = \begin{bmatrix} \cos\theta & \sin\theta \\ -\sin\theta & \cos\theta \end{bmatrix} = A^{-1}.$$

One can prove this by evaluating the product AA^T to show that it equals the identity matrix I:

$$AA^T = \begin{bmatrix} \cos\theta & -\sin\theta \\ \sin\theta & \cos\theta \end{bmatrix} \begin{bmatrix} \cos\theta & \sin\theta \\ -\sin\theta & \cos\theta \end{bmatrix}$$

$$= \begin{bmatrix} \cos^2\theta + \sin^2\theta & 0 \\ 0 & \cos^2 + \sin^2\theta \end{bmatrix}$$

$$= \begin{bmatrix} 1 & 0 \\ 0 & 1 \end{bmatrix} = I$$

which confirms that $A^T = A^{-1}$.

It goes without saying that the identity matrix is orthogonal:

$$II^T = \begin{bmatrix} 1 & 0 & 0 \\ 0 & 1 & 0 \\ 0 & 0 & 1 \end{bmatrix} \begin{bmatrix} 1 & 0 & 0 \\ 0 & 1 & 0 \\ 0 & 0 & 1 \end{bmatrix}$$

$$= \begin{bmatrix} 1 & 0 & 0 \\ 0 & 1 & 0 \\ 0 & 0 & 1 \end{bmatrix}.$$

Naturally, a complex matrix can also be orthogonal. For example, the following matrix satisfies the definition for orthogonality:

$$A = \begin{bmatrix} i & \sqrt{2} \\ -\sqrt{2} & i \end{bmatrix}$$

$$AA^T = \begin{bmatrix} i & \sqrt{2} \\ -\sqrt{2} & i \end{bmatrix} \begin{bmatrix} i & -\sqrt{2} \\ \sqrt{2} & i \end{bmatrix}$$

$$= \begin{bmatrix} 1 & 0 \\ 0 & 1 \end{bmatrix}.$$

3.2.20 Unitary Matrix

A *unitary matrix* is a complex square matrix whose conjugate transpose is also its inverse:

$$\overline{A^T}A = I.$$

The following matrix is unitary

$$A = \begin{bmatrix} \frac{1}{\sqrt{2}} & \frac{1}{\sqrt{2}} & 0 \\ -\frac{i}{\sqrt{2}} & \frac{i}{\sqrt{2}} & 0 \\ 0 & 0 & i \end{bmatrix}$$

because

$$A^{\mathrm{T}} = \begin{bmatrix} \frac{1}{\sqrt{2}} & -\frac{i}{\sqrt{2}} & 0 \\ \frac{1}{\sqrt{2}} & \frac{i}{\sqrt{2}} & 0 \\ 0 & 0 & i \end{bmatrix}$$

$$\overline{A^{\mathrm{T}}} = \begin{bmatrix} \frac{1}{\sqrt{2}} & \frac{i}{\sqrt{2}} & 0 \\ \frac{1}{\sqrt{2}} & -\frac{i}{\sqrt{2}} & 0 \\ 0 & 0 & -i \end{bmatrix}$$

$$\overline{A^{\mathrm{T}}} A = \begin{bmatrix} \frac{1}{\sqrt{2}} & \frac{i}{\sqrt{2}} & 0 \\ \frac{1}{\sqrt{2}} & -\frac{i}{\sqrt{2}} & 0 \\ 0 & 0 & -i \end{bmatrix} \begin{bmatrix} \frac{1}{\sqrt{2}} & \frac{1}{\sqrt{2}} & 0 \\ -\frac{i}{\sqrt{2}} & \frac{i}{\sqrt{2}} & 0 \\ 0 & 0 & i \end{bmatrix}$$

$$= \begin{bmatrix} 1 & 0 & 0 \\ 0 & 1 & 0 \\ 0 & 0 & 1 \end{bmatrix}.$$

A unitary matrix is also normal, i.e. $\overline{A^{\mathrm{T}}} A = A \overline{A^{\mathrm{T}}}$.

3.3 Eigenvectors and Eigenvalues

3.3.1 Real Eigenvectors and Eigenvalues

The German mathematician David Hilbert (1862–1943), is associated with coining the terms *eigenwert* and *eigenvektor* in 1904. The English translation of *eigen* is *proper* or *characteristic*, where an eigenvector is a special vector associated with a matrix. Let's begin by defining an eigenvector.

When the following transform A acts upon different vectors, generally, it creates a vector with a new orientation and length.

$$A = \begin{bmatrix} 4 & 1 \\ 1 & 4 \end{bmatrix}.$$

For example, if $\mathbf{v} = [2 \quad 3]^{\mathrm{T}}$, then

$$A\mathbf{v} = \begin{bmatrix} 4 & 1 \\ 1 & 4 \end{bmatrix} \begin{bmatrix} 2 \\ 3 \end{bmatrix} = \begin{bmatrix} 11 \\ 14 \end{bmatrix}$$

which has a different orientation and length to $\mathbf{v}$.

Similarly, if $\mathbf{v} = [1 \quad -3]^{\mathrm{T}}$, then

$$A\mathbf{v} = \begin{bmatrix} 4 & 1 \\ 1 & 4 \end{bmatrix} \begin{bmatrix} 1 \\ -3 \end{bmatrix} = \begin{bmatrix} 1 \\ -11 \end{bmatrix}$$

which also has a different orientation and length to $\mathbf{v}$.

However, if $\mathbf{v} = [2 \quad 2]^{\mathrm{T}}$, then

$$A\mathbf{v} = \begin{bmatrix} 4 & 1 \\ 1 & 4 \end{bmatrix} \begin{bmatrix} 2 \\ 2 \end{bmatrix} = \begin{bmatrix} 10 \\ 10 \end{bmatrix}$$

which has the same orientation as $\mathbf{v}$, and has been stretched by a factor of 5.

Similarly, if $\mathbf{v} = [2 \quad -2]^{\mathrm{T}}$, then

$$A\mathbf{v} = \begin{bmatrix} 4 & 1 \\ 1 & 4 \end{bmatrix} \begin{bmatrix} 2 \\ -2 \end{bmatrix} = \begin{bmatrix} 6 \\ -6 \end{bmatrix}$$

which also has the same orientation as $\mathbf{v}$, and has been stretched by a factor of 3.

The two vectors, $[2 \quad 2]^{\mathrm{T}}$ and $[2 \quad -2]^{\mathrm{T}}$ are the eigenvectors, and the two stretching factors, 5 and 3, are the associated eigenvalues.

We identify an eigenvector and its eigenvalue as follows. Given a square matrix A, then a non-zero vector $\mathbf{v}$ is an eigenvector, and a scalar λ is the corresponding eigenvalue if

$$A\mathbf{v} = \lambda\mathbf{v}.$$

The equation that determines the existence of any eigenvectors is called the *characteristic equation* of a square matrix, and is given by

$$|A - \lambda I| = 0. \tag{3.4}$$

Let's derive the characteristic equation (3.4).

Consider the following matrix A and vector $\mathbf{v}$:

$$A = \begin{bmatrix} a & b \\ c & d \end{bmatrix}, \quad \mathbf{v} = \begin{bmatrix} x \\ y \end{bmatrix}.$$

If **v** is an eigenvector of A, and λ its associated eigenvalue, then

$$A\mathbf{v} = \lambda\mathbf{v}$$
$$= \lambda I \mathbf{v}$$
$$(A - \lambda I)\mathbf{v} = \mathbf{0}$$
$$\begin{bmatrix} a - \lambda & b \\ c & d - \lambda \end{bmatrix}\begin{bmatrix} x \\ y \end{bmatrix} = \begin{bmatrix} 0 \\ 0 \end{bmatrix}.$$

For a non-zero eigenvector $[x \ \ y]^{\mathrm{T}}$ to exist, we must have

$$\begin{vmatrix} a - \lambda & b \\ c & d - \lambda \end{vmatrix} = 0.$$

Let's use this on the matrix A:

$$A = \begin{bmatrix} 4 & 1 \\ 1 & 4 \end{bmatrix}$$

then

$$\begin{vmatrix} 4 - \lambda & 1 \\ 1 & 4 - \lambda \end{vmatrix} = 0$$
$$(4 - \lambda)^2 - 1 = 0$$
$$\lambda^2 - 8\lambda + 16 - 1 = 0$$
$$\lambda^2 - 8\lambda + 15 = 0$$
$$(\lambda - 5)(\lambda - 3) = 0.$$

Thus $\lambda = 5$ and $\lambda = 3$, are the two eigenvalues, which make the matrix singular, due to its zero determinant. Figure 3.2 shows the graph of the quadratic in λ, with the two roots.

Fig. 3.2 Graph of $y = \lambda^2 - 8\lambda + 15$

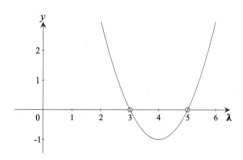

To discover the associated eigenvectors, we substitute the values of λ in

$$\begin{bmatrix} 4-\lambda & 1 \\ 1 & 4-\lambda \end{bmatrix} \begin{bmatrix} x \\ y \end{bmatrix} = \begin{bmatrix} 0 \\ 0 \end{bmatrix}.$$

Let's start with $\lambda = 5$:

$$\begin{bmatrix} -1 & 1 \\ 1 & -1 \end{bmatrix} \begin{bmatrix} x \\ y \end{bmatrix} = \begin{bmatrix} 0 \\ 0 \end{bmatrix}$$

which represents the equation $y = x$ or the vector $[1 \quad 1]^T$. Note that the equal and opposite vector $[-1 \quad -1]^T$ is also a solution.

Next, we substitute $\lambda = 3$:

$$\begin{bmatrix} 1 & 1 \\ 1 & 1 \end{bmatrix} \begin{bmatrix} x \\ y \end{bmatrix} = \begin{bmatrix} 0 \\ 0 \end{bmatrix}$$

which represents the equation $y = -x$ or the vector $[1 \quad -1]^T$. Note that the equal and opposite vector $[-1 \quad 1]^T$ is also a solution. Thus we have discovered that the two eigenvectors are $[1 \quad 1]^T$ and $[1 \quad -1]^T$, with their mirror vectors, and their respective eigenvalues $\lambda = 5$ and $\lambda = 3$, as predicted.

Figure 3.3 shows points on the unit circle, and arrows pointing to their positions after the matrix operation. The blue arrows align with the eigenvectors, and their eigenvalues, are the lengths relative to the origin.

The characteristic equation in the above example is a quadratic in λ, as it came from a 2×2 matrix. Increasing the order of the matrix, increases the order of the polynomial, which sometimes makes it necessary to employ software to evaluate the eigenvalues and eigenvectors. Both eigenvalues are real positive numbers, but they can take on a wide range of values, including complex.

A 2×2 matrix A will normally give rise to two eigenvalues and two eigenvectors, but in the case of $A = I$, (the identity matrix), we have

Fig. 3.3 How points on the unit circle move to their transformed positions

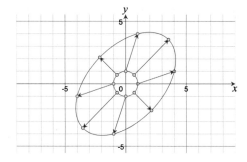

$$A\mathbf{v} = \lambda\mathbf{v}$$
$$I\mathbf{v} = \lambda I\mathbf{v}$$
$$\mathbf{v} = \lambda\mathbf{v}$$

and $\lambda = \pm 1$. Furthermore, all vectors must be eigenvectors.

3.3.2 Complex Eigenvectors and Eigenvalues

The characteristic equation may have real or complex roots, and if they are complex, there are no real eigenvectors. For example, let's examine the following matrix:

$$A = \begin{bmatrix} 6 & -4 \\ 8 & -2 \end{bmatrix}.$$

The characteristic equation is

$$\begin{vmatrix} 6 - \lambda & -4 \\ 8 & -2 - \lambda \end{vmatrix} = (6 - \lambda)(-2 - \lambda) - (-4)8 = 0$$
$$0 = -12 - 6\lambda + 2\lambda + \lambda^2 + 32$$
$$= \lambda^2 - 4\lambda + 20$$

which has roots given by

$$\lambda = \frac{4 \pm \sqrt{16 - 80}}{2} = 2 \pm 4i$$

and form a complex conjugate pair. The graph of this quadratic is shown in Fig. 3.4.

To find the corresponding eigenvectors, we substitute each eigenvalue into $A\mathbf{v} = \lambda\mathbf{v}$:

Fig. 3.4 Graph of $y = \lambda^2 - 4\lambda + 20$

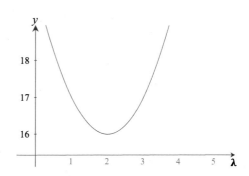

$$\mathbf{A}\mathbf{v} = (2 + 4i)\mathbf{v}$$

$$\begin{bmatrix} 6 & -4 \\ 8 & -2 \end{bmatrix} \begin{bmatrix} x \\ y \end{bmatrix} = (2 + 4i) \begin{bmatrix} 1 & 0 \\ 0 & 1 \end{bmatrix} \begin{bmatrix} x \\ y \end{bmatrix}$$

$$\begin{bmatrix} 4 - 4i & -4 \\ 8 & -(4 + 4i) \end{bmatrix} \begin{bmatrix} x \\ y \end{bmatrix} = \begin{bmatrix} 0 \\ 0 \end{bmatrix}$$

which provides two equations:

$$(4 - 4i)x - 4y = 0$$
$$8x - (4 + 4i)y = 0.$$

However, the determinant of the associated matrix is zero:

$$\begin{vmatrix} 4 - 4i & -4 \\ 8 & -4 - 4i \end{vmatrix} = 0$$

and either equation can be used:

$$(4 - 4i)x - 4y = (1 - i)x - y = 0$$
$$y = (1 - i)x. \tag{3.5}$$

A cunning way to simplify (3.5) is to let $k = x$:

$$\mathbf{v} = \begin{bmatrix} x \\ y \end{bmatrix} = \begin{bmatrix} k \\ k(1 - i) \end{bmatrix} = k \begin{bmatrix} 1 \\ 1 - i \end{bmatrix}$$

where k is an arbitrary scalar.

Next, we take the second eigenvalue $2 - 4i$, and show that the corresponding eigenvector is

$$\mathbf{v} = k \begin{bmatrix} 1 \\ 1 + i \end{bmatrix}.$$

We have shown that the eigenvalues and eigenvectors are

$$\lambda = 2 + 4i, \quad \mathbf{v} = \begin{bmatrix} 1 \\ 1 - i \end{bmatrix}$$

$$\lambda = 2 - 4i, \quad \mathbf{v} = \begin{bmatrix} 1 \\ 1 + i \end{bmatrix}$$

now let's show that $A\mathbf{v} = \lambda\mathbf{v}$ for both pairs:

$$A\mathbf{v} = \begin{bmatrix} 6 & -4 \\ 8 & -2 \end{bmatrix} \begin{bmatrix} 1 \\ 1-i \end{bmatrix} = \begin{bmatrix} 2+4i \\ 6+2i \end{bmatrix}$$

$$\lambda\mathbf{v} = (2+4i) \begin{bmatrix} 1 \\ 1-i \end{bmatrix} = \begin{bmatrix} 2+4i \\ (2+4i)(1-i) \end{bmatrix} = \begin{bmatrix} 2+4i \\ 6+2i \end{bmatrix}$$

$$A\mathbf{v} = \begin{bmatrix} 6 & -4 \\ 8 & -2 \end{bmatrix} \begin{bmatrix} 1 \\ 1+i \end{bmatrix} = \begin{bmatrix} 2-4i \\ 6-2i \end{bmatrix}$$

$$\lambda\mathbf{v} = (2-4i) \begin{bmatrix} 1 \\ 1+i \end{bmatrix} = \begin{bmatrix} 2-4i \\ (2-4i)(1+i) \end{bmatrix} = \begin{bmatrix} 2-4i \\ 6-2i \end{bmatrix}.$$

In fact, if the square matrix A has real entries and complex eigenvalues, they will always occur in complex conjugate pairs due to the form of the equation giving the roots. Furthermore, if

$$A\mathbf{v} = \lambda\mathbf{v}$$

where $\mathbf{v}$ is a non-zero eigenvector, then taking complex conjugates:

$$\overline{A\mathbf{v}} = \overline{\lambda}\overline{\mathbf{v}}$$

and as A has real entries, $\overline{A} = A$, which means

$$A\overline{\mathbf{v}} = \overline{\lambda}\overline{\mathbf{v}}$$

i.e. the eigenvalues and eigenvectors come in complex conjugate pairs.

3.3.3 Eigenvectors of a Rotation Matrix

We would not expect a rotation transform to have any specific eigenvector, as this would imply that it shows a rotational preference to certain points. Let's explore this transform to see how the characteristic equation behaves.

$$A = \begin{bmatrix} \cos\beta & -\sin\beta \\ \sin\beta & \cos\beta \end{bmatrix}$$

where β is the angle of rotation.

The characteristic equation is

$$\begin{vmatrix} \cos(\beta) - \lambda & -\sin\beta \\ \sin\beta & \cos(\beta) - \lambda \end{vmatrix} = 0 \tag{3.6}$$

therefore,

$$\left(\cos(\beta) - \lambda\right)^2 + \sin^2 \beta = 0$$
$$\lambda^2 - 2\lambda \cos \beta + \cos^2 \beta + \sin^2 \beta = 0$$
$$\lambda^2 - 2\lambda \cos \beta + 1 = 0. \qquad (3.7)$$

Equation (3.7) is a quadratic in λ and solved using

$$\lambda = \frac{-b \pm \sqrt{b^2 - 4ac}}{2a}$$

where $a = 1, b = -2\cos \beta, c = 1$:

$$\lambda = \frac{2\cos \beta \pm \sqrt{4\cos^2 \beta - 4}}{2}$$
$$= \cos \beta \pm \sqrt{\cos^2 \beta - 1}$$
$$= \cos \beta \pm \sqrt{- \sin^2 \beta}$$
$$\lambda_1 = \cos \beta + i \sin \beta$$
$$\lambda_2 = \cos \beta - i \sin \beta.$$

A problem with trigonometric functions is that they have multiple zero and non-zero values. So let's test for these conditions.

$$(A - \lambda I)\mathbf{v} = \mathbf{0}$$
$$\begin{bmatrix} \cos(\beta) - \lambda & -\sin \beta \\ \sin \beta & \cos(\beta) - \lambda \end{bmatrix} \begin{bmatrix} x \\ y \end{bmatrix} = \begin{bmatrix} 0 \\ 0 \end{bmatrix}.$$

When $\beta = 0$ or π, then $\lambda = 1$, and which implies that each non-zero vector $\mathbf{v}$ is an eigenvector.

Now let's find the eigenvectors for the eigenvalues when $\beta \neq 0$, starting with $\lambda_1 = \cos \beta + i \sin \beta$.

Using the characteristic equation:

$$|A - \lambda I| = 0$$
$$\begin{vmatrix} \cos(\beta) - \lambda_1 & -\sin \beta \\ \sin \beta & \cos(\beta) - \lambda_1 \end{vmatrix} = 0$$
$$\begin{vmatrix} -i \sin \beta & -\sin \beta \\ \sin \beta & -i \sin \beta \end{vmatrix} = -\sin^2 \beta + \sin^2 \beta = 0$$

which means that the two associated equations are identical, and either can be used:

$$-ix \sin \beta - y \sin \beta = ix - y = 0$$
$$y = ix.$$

This is resolved by letting $x = k$:

$$\mathbf{v} = \begin{bmatrix} x \\ y \end{bmatrix} = \begin{bmatrix} k \\ ki \end{bmatrix} = k \begin{bmatrix} 1 \\ i \end{bmatrix}$$

where k is an arbitrary scalar.

The corresponding complex eigenvectors are

$$\mathbf{v}_1 = \begin{bmatrix} 1 \\ i \end{bmatrix}$$

$$\mathbf{v}_2 = \begin{bmatrix} 1 \\ -i \end{bmatrix}.$$

To summarise:

When $\beta = 0$ then $\lambda_1 = \lambda_2 = 1$, and when $\beta = \pi$ then $\lambda_1 = \lambda_2 = -1$, and all non-zero vectors are eigenvectors.

When $\beta \neq 0$ then $\lambda_1 = \cos \beta + i \sin \beta$ with the eigenvector $[1 \quad i]^T$, $\lambda_2 = \cos \beta - i \sin \beta$ with the eigenvector $[1 \quad -i]^T$.

3.4 Representing a Complex Number as a Matrix

To illustrate how a complex number is represented by a matrix, we exploit the fact that a complex number can be represented as an ordered pair or a vector. We also bear in mind that the product of two matrices must generate the same answer as their algebraic form. Let's begin by multiplying two complex numbers z_1 and z_2 together to separate their real and imaginary parts.

$$z_1 = a_1 + b_1 i$$
$$z_2 = a_2 + b_2 i$$
$$z_1 z_2 = (a_1 + b_1 i)(a_2 + b_2 i)$$
$$= a_1 a_2 - b_1 b_2 + (b_1 a_2 + a_1 b_2)i.$$

We now represent the product $z_1 z_2$ and z_2 as column vectors, and introduce a 2×2 matrix with unknown terms to represent z_1:

$$\begin{bmatrix} a_1 a_2 - b_1 b_2 \\ b_1 a_2 + a_1 b_2 \end{bmatrix} = \begin{bmatrix} A & B \\ C & D \end{bmatrix} \begin{bmatrix} a_2 \\ b_2 \end{bmatrix}. \tag{3.8}$$

It is clear from (3.8) that $A a_2 + B b_2$ must equal $a_1 a_2 - b_1 b_2$, which makes $A = a_1$ and $B = -b_2$. Similarly, $C a_2 + D b_2$ must equal $b_1 a_2 + a_1 b_2$, which makes $C = b_1$ and $D = a_1$. Therefore,

$$\begin{bmatrix} a_1 a_2 - b_1 b_2 \\ b_1 a_2 + a_1 b_2 \end{bmatrix} = \begin{bmatrix} a_1 & -b_1 \\ b_1 & a_1 \end{bmatrix} \begin{bmatrix} a_2 \\ b_2 \end{bmatrix}. \tag{3.9}$$

The 2×2 matrix in (3.9) is equivalent to the complex number z_1:

$$a_1 + b_1 i \equiv \begin{bmatrix} a_1 & -b_1 \\ b_1 & a_1 \end{bmatrix}. \tag{3.10}$$

For example, the complex number $2 + 3i$ is equivalent to the matrix

$$2 + 3i \equiv \begin{bmatrix} 2 & -3 \\ 3 & 2 \end{bmatrix}.$$

Let's confirm this further by evaluating the product of two complex numbers algebraically and by using matrices.

$$z_1 = 2 + 3i$$
$$z_2 = 4 + 5i$$
$$z_1 z_2 = (2 + 3i)(4 + 5i)$$
$$= 8 - 15 + 12i + 10i$$
$$= -7 + 22i.$$

Now using vectors and matrices, where

$$4 + 5i \equiv [4 \quad 5]^{\mathrm{T}}$$
$$2 + 3i \equiv \begin{bmatrix} 2 & -3 \\ 3 & 2 \end{bmatrix}$$

therefore,

$$z_1 z_2 \equiv \begin{bmatrix} 2 & -3 \\ 3 & 2 \end{bmatrix} \begin{bmatrix} 4 \\ 5 \end{bmatrix}$$
$$\equiv \begin{bmatrix} 8 - 15 \\ 12 + 10 \end{bmatrix}$$
$$\equiv \begin{bmatrix} -7 \\ 22 \end{bmatrix} \equiv -7 + 22i.$$

Let's explore (3.10) further. The matrix for 1 is

$$1 \equiv \begin{bmatrix} 1 & 0 \\ 0 & 1 \end{bmatrix}$$

which is the identity matrix and leaves any matrix it multiplies untouched.

The matrix for the imaginary unit i is

$$i \equiv \begin{bmatrix} 0 & -1 \\ 1 & 0 \end{bmatrix}. \tag{3.11}$$

If this is correct, then the square of the matrix must equal another matrix whose effect is to multiply by -1:

$$\begin{bmatrix} 0 & -1 \\ 1 & 0 \end{bmatrix}\begin{bmatrix} 0 & -1 \\ 1 & 0 \end{bmatrix} = \begin{bmatrix} -1 & 0 \\ 0 & -1 \end{bmatrix}$$

$$= -1\begin{bmatrix} 1 & 0 \\ 0 & 1 \end{bmatrix}$$

and confirms that (3.11) is correct.

Having computed the matrix for i, its inverse is

$$i^{-1} \equiv \begin{bmatrix} 0 & 1 \\ -1 & 0 \end{bmatrix}$$

because

$$\frac{1}{a+bi} = \frac{(a-bi)}{(a-bi)}\frac{1}{(a+bi)}$$

$$= \frac{a-bi}{a^2+b^2}$$

and as $x = 0$ and $y = 1$ then

$$\frac{1}{0+i} = 0 - i$$

$$i^{-1} \equiv \begin{bmatrix} 0 & 1 \\ -1 & 0 \end{bmatrix}.$$

Multiplying the matrices for i and i^{-1} must give the identity matrix:

$$ii^{-1} \equiv \begin{bmatrix} 0 & -1 \\ 1 & 0 \end{bmatrix} \begin{bmatrix} 0 & 1 \\ -1 & 0 \end{bmatrix}$$

$$\equiv \begin{bmatrix} 1 & 0 \\ 0 & 1 \end{bmatrix}.$$

Finally, let $x = \cos\theta$, and $y = \sin\theta$, to represent the polar form $e^{i\theta} = \cos\theta + i\sin\theta$:

$$a + bi \equiv \begin{bmatrix} x & -y \\ y & x \end{bmatrix}$$

$$e^{i\theta} = \cos\theta + i\sin\theta \equiv \begin{bmatrix} \cos\theta & -\sin\theta \\ \sin\theta & \cos\theta \end{bmatrix}. \qquad (3.12)$$

The 2×2 matrix in (3.12) is known as a rotation matrix, and rotates a position vector θ about the origin, which is the effect of multiplying a complex number by $\cos\theta + i\sin\theta$.

Now let's compute the determinant of the matrix representing a complex number:

$$z = a + bi$$

$$a + bi \equiv \begin{bmatrix} a & -b \\ b & a \end{bmatrix}$$

$$\det \begin{bmatrix} a & -b \\ b & a \end{bmatrix} = a^2 + b^2$$

$$= |z|^2$$

which is the modulus, squared.

3.5   Complex Algebra Using Matrices

Now that we have the following matrices:

$$a + bi \equiv \begin{bmatrix} a & -b \\ b & a \end{bmatrix}$$

$$i \equiv \begin{bmatrix} 0 & -1 \\ 1 & 0 \end{bmatrix}$$

$$\cos\theta + i\sin\theta \equiv \begin{bmatrix} \cos\theta & -\sin\theta \\ \sin\theta & \cos\theta \end{bmatrix}$$

we can undertake complex algebra without using i directly.

For example, let's compute the product $(2 + 4i)(3 + 5i)$ using matrices.

$$(2 + 4i)(3 + 5i) \equiv \begin{bmatrix} 2 & -4 \\ 4 & 2 \end{bmatrix} \begin{bmatrix} 3 & -5 \\ 5 & 3 \end{bmatrix}$$

$$\equiv \begin{bmatrix} 6 - 20 & -10 - 12 \\ 12 + 10 & -20 + 6 \end{bmatrix}$$

$$\equiv \begin{bmatrix} -14 & -22 \\ 22 & -14 \end{bmatrix}$$

$$= -14 + 22i.$$

Now let's compute the product $i(3 \cos \theta + 2i \sin \theta)$ using matrices.

$$i(3 \cos \theta + 2i \sin \theta) \equiv \begin{bmatrix} 0 & -1 \\ 1 & 0 \end{bmatrix} \begin{bmatrix} 3 \cos \theta & -2 \sin \theta \\ 2 \sin \theta & 3 \cos \theta \end{bmatrix}$$

$$\equiv \begin{bmatrix} -2 \sin \theta & -3 \cos \theta \\ 3 \cos \theta & -2 \sin \theta \end{bmatrix}$$

$$= -2 \sin \theta + 3i \cos \theta.$$

Finally, the product $(2 + 3i)(3 \cos \theta + 2i \sin \theta)$ using matrices.

$$(2 + 3i)(3 \cos \theta + 2i \sin \theta) \equiv \begin{bmatrix} 2 & -3 \\ 3 & 2 \end{bmatrix} \begin{bmatrix} 3 \cos \theta & -2 \sin \theta \\ 2 \sin \theta & 3 \cos \theta \end{bmatrix}$$

$$\equiv \begin{bmatrix} 6 \cos \theta - 6 \sin \theta & -4 \sin \theta - 9 \cos \theta \\ 9 \cos \theta + 4 \sin \theta & -6 \sin \theta + 6 \cos \theta \end{bmatrix}$$

$$= 6(\cos \theta - \sin \theta) + (9 \cos \theta + 4 \sin \theta)i.$$

3.6 Complex Vectors

3.6.1 Cartesian Vector Space

Cartesian vectors are normally used to represent physical quantities possessing a magnitude and direction. The mathematical construct is a linear combination of the basis vectors, which for $\mathbb{R}^n$ are

$$\mathbf{e}_1 = (1, \ 0, \ 0, \ldots, 0)$$
$$\mathbf{e}_2 = (0, \ 1, \ 0, \ldots, 0)$$
$$\vdots$$
$$\mathbf{e}_n = (0, \ 0, \ 0, \ldots, 1).$$

Any vector $\mathbf{v}$ is a linear combination of $\mathbf{e}_1, \mathbf{e}_2, \ldots, \mathbf{e}_n$:

$$\mathbf{v} = r_1\mathbf{e}_1 + r_2\mathbf{e}_2 + \cdots + r_n\mathbf{e}_n \quad \text{where} \quad \{r_1, r_2, \ldots, r_n\} \in \mathbb{R}$$

and

$$\mathbf{v} = [r_1, \ r_2, \ldots, r_n].$$

3.6.2   Complex Vector Space

The complex vector space $\mathbb{C}^n$ employs the same standard basis, but substitutes complex numbers, instead of real scalars:

$$\mathbf{v} = c_1\mathbf{e}_1 + c_2\mathbf{e}_2 + \cdots + c_n\mathbf{e}_n, \quad \{c_1, c_2, \ldots, c_n\} \in \mathbb{C}$$

and

$$\mathbf{v} = [c_1, c_2, \ldots, c_n].$$

For example, $\mathbf{v}$ is a complex vector in $\mathbb{C}^3$:

$$\mathbf{v} = \begin{bmatrix} 2 + 3i \\ 1 - 4i \\ 3 + 5i \end{bmatrix}.$$

Complex vectors in $\mathbb{C}^n$ can be scaled, added and subtracted, just like real vectors in $\mathbb{R}^n$. For example,

$$\mathbf{u} = \begin{bmatrix} 2 + 3i \\ 1 - 4i \\ 3 + 5i \end{bmatrix}, \quad \mathbf{v} = \begin{bmatrix} 4 + 4i \\ 1 + 4i \\ 5 + 3i \end{bmatrix}$$

$$\mathbf{u} + \mathbf{v} = \begin{bmatrix} 6 + 7i \\ 2 \\ 8 + 8i \end{bmatrix}$$

and

$$(2 + i)\mathbf{u} = (2 + i) \begin{bmatrix} 2 + 3i \\ 1 - 4i \\ 3 + 5i \end{bmatrix}$$

$$= \begin{bmatrix} (2+i)(2+3i) \\ (2+i)(1-4i) \\ (2+i)(3+5i) \end{bmatrix}$$

$$= \begin{bmatrix} 1+8i \\ 6-7i \\ 1+13i \end{bmatrix}.$$

3.6.3 Inner Product in $\mathbb{R}^n$

I have included the *inner product* in $\mathbb{R}^n$, also called the *Euclidean inner product* and the *dot product*, to highlight the similarity between $\mathbb{R}^n$ and $\mathbb{C}^n$.

Given two column vectors **u** and **v**:

$$\mathbf{u} = \begin{bmatrix} u_1 \\ u_2 \\ \vdots \\ u_n \end{bmatrix}, \quad \mathbf{v} = \begin{bmatrix} v_1 \\ v_2 \\ \vdots \\ v_n \end{bmatrix}$$

their inner product is a scalar:

$$\mathbf{u} \cdot \mathbf{v} = \mathbf{u}^\mathsf{T} \mathbf{v} = [u_1, u_2, \dots, u_n] \begin{bmatrix} v_1 \\ v_2 \\ \vdots \\ v_n \end{bmatrix}$$

$$= \sum_{i=1}^{n} u_i v_i = u_1 v_1 + u_2 v_2 + \cdots + u_n v_n$$

where $\{u_1, u_2, \dots, u_n, v_1, v_2, \dots, v_n\} \in \mathbb{R}$.

The inner product also reveals the angle between **u** and **v**:

$$\cos \theta = \frac{\mathbf{u} \cdot \mathbf{v}}{|\mathbf{u}| \, |\mathbf{v}|}$$

where $|\mathbf{u}|$ is the length of **u**, which also equals the *Euclidean norm* $\|\mathbf{u}\|$:

$$\|\mathbf{u}\| = \sqrt{\mathbf{u} \cdot \mathbf{u}}$$

$$= \sqrt{\sum_{i=1}^{n} u_i^2} = \sqrt{u_1^2 + u_2^2 + \cdots + u_n^2}.$$

If $\mathbf{u}$ and $\mathbf{v}$ are regarded as position vectors for two points, the distance between the vectors, or points, is $d(\mathbf{u}, \mathbf{v})$:

$$
\begin{aligned}
d(\mathbf{u}, \ \mathbf{v}) &= |\mathbf{u} - \mathbf{v}| \\
&= \sqrt{(\mathbf{u} - \mathbf{v}) \cdot (\mathbf{u} - \mathbf{v})} \\
&= \sqrt{\sum_{i=1}^{n}(u_i - v_i)^2} = \sqrt{(u_1 - v_1)^2 + (u_2 - v_2)^2 + \cdots + (u_n - v_n)^2}.
\end{aligned}
$$

3.6.4   Inner Product in $\mathbb{C}^n$

The *inner product* in $\mathbb{C}^n$ is similar to the inner product in $\mathbb{R}^n$, but requires the complex conjugate of one of the vectors:

$$
\mathbf{u} = \begin{bmatrix} u_1 \\ u_2 \\ \vdots \\ u_n \end{bmatrix}, \quad
\mathbf{v} = \begin{bmatrix} v_1 \\ v_2 \\ \vdots \\ v_n \end{bmatrix}
$$

their inner product is:

$$
\mathbf{u} \cdot \mathbf{v} = \mathbf{u}^T \bar{\mathbf{v}} = [u_1, u_2, \ldots, u_n] \begin{bmatrix} \bar{v}_1 \\ \bar{v}_2 \\ \vdots \\ \bar{v}_n \end{bmatrix}
$$

$$
= \sum_{i=1}^{n} u_i \bar{v}_i = u_1 \bar{v}_1 + u_2 \bar{v}_2 + \cdots + u_n \bar{v}_n
$$

where $\{u_1, u_2, \ldots, u_n, v_1, v_2, \ldots, v_n\} \in \mathbb{C}$.

For example,

$$
\mathbf{u} = \begin{bmatrix} 1 + 2i \\ 2 - 3i \\ 4 + 3i \end{bmatrix}, \quad
\mathbf{v} = \begin{bmatrix} 2 - i \\ 3 + 3i \\ 1 + 2i \end{bmatrix}
$$

$$
\mathbf{u} \cdot \mathbf{v} = \mathbf{u}^T \bar{\mathbf{v}} = [1 + 2i, \ 2 - 3i, \ 4 + 3i] \begin{bmatrix} 2 + i \\ 3 - 3i \\ 1 - 2i \end{bmatrix}
$$

$$= (1 + 2i)(2 + i) + (2 - 3i)(3 - 3i) + (4 + 3i)(1 - 2i)$$
$$= 5i - 3 - 15i + 10 - 5i$$
$$= 7 - 15i.$$

The Euclidean norm of $\mathbf{u}$ in $\mathbb{C}^n$ is

$$\|\mathbf{u}\| = \sqrt{\mathbf{u} \cdot \mathbf{u}}$$
$$= \sqrt{|u_1|^2 + |u_2|^2 + \cdots + |u_n|^2}.$$

For example,

$$\mathbf{u} = [1 + 2i, \ 2 - 3i, \ 4 + 3i]$$
$$\|\mathbf{u}\| = \sqrt{|1 + 2i|^2 + |2 - 3i|^2 + |4 + 3i|^2}$$
$$= \sqrt{(1^2 + 2^2) + (2^2 + 3^2) + (4^2 + 3^2)}$$
$$= \sqrt{5 + 13 + 25} = \sqrt{43}.$$

The Euclidean distance between $\mathbf{u}$ and $\mathbf{v}$ is $d(\mathbf{u}, \mathbf{v})$:

$$d(\mathbf{u}, \ \mathbf{v}) = |\mathbf{u} - \mathbf{v}|$$
$$= \sqrt{(\mathbf{u} - \mathbf{v}) \cdot (\mathbf{u} - \mathbf{v})}$$
$$= \sqrt{\sum_{i=1}^{n} |u_i - v_i|^2} = \sqrt{|u_1 - v_1|^2 + |u_2 - v_2|^2 + \cdots + |u_n - v_n|^2}.$$

For example,

$$\mathbf{u} = [1 + 2i, \ 2 - 3i, \ 4 + 3i]$$
$$\mathbf{v} = [2 - i, \ 3 + 3i, \ 1 + 2i]$$
$$d(\mathbf{u}, \ \mathbf{v}) = |\mathbf{u} - \mathbf{v}|$$
$$= |(-1 + 3i), \ (-1 - 6i), \ (3 + i)|$$
$$= \sqrt{1^2 + 3^2 + 1^2 + 6^2 + 3^2 + 1^2}$$
$$= \sqrt{1 + 9 + 1 + 36 + 9 + 1} = \sqrt{57}.$$

3.6.5 *Outer Product in* $\mathbb{R}^n$

Given two column vectors $\mathbf{u}$ and $\mathbf{v}$:

$$\mathbf{u} = \begin{bmatrix} u_1 \\ u_2 \\ \vdots \\ u_n \end{bmatrix}, \qquad \mathbf{v} = \begin{bmatrix} v_1 \\ v_2 \\ \vdots \\ v_n \end{bmatrix}$$

where $\{u_1, u_2, \ldots, u_n, v_1, v_2, \ldots, v_n\} \in \mathbb{R}$, their outer product is a matrix:

$$A = \mathbf{u} \otimes \mathbf{v} = \mathbf{u}\mathbf{v}^{\mathrm{T}}$$

$$= \begin{bmatrix} u_1 \\ u_2 \\ \vdots \\ u_n \end{bmatrix} [v_1, v_2, \cdots, v_n]$$

$$= \begin{bmatrix} u_1 v_1 & u_1 v_2 & \cdots & u_1 v_n \\ u_2 v_1 & u_2 v_2 & \cdots & u_2 v_n \\ \vdots & \vdots & \ddots & \vdots \\ u_n v_1 & u_n v_2 & \cdots & u_n v_n \end{bmatrix}.$$

For example,

$$\mathbf{u} = \begin{bmatrix} 2 \\ 3 \\ 4 \end{bmatrix}, \qquad \mathbf{v} = \begin{bmatrix} 5 \\ 6 \\ 7 \end{bmatrix}$$

$$\mathbf{u} \otimes \mathbf{v} = \mathbf{u}\mathbf{v}^{\mathrm{T}} = \begin{bmatrix} 2 \\ 3 \\ 4 \end{bmatrix} [5 \quad 6 \quad 7]$$

$$= \begin{bmatrix} 10 & 12 & 14 \\ 15 & 18 & 21 \\ 20 & 24 & 28 \end{bmatrix}.$$

3.6.6 Outer Product in $\mathbb{C}^n$

The only difference between the outer product in $\mathbb{R}^n$ and $\mathbb{C}^n$ is that the transposed vector is also the complex conjugate. Given two column complex vectors $\mathbf{u}$ and $\mathbf{v}$:

$$\mathbf{u} = \begin{bmatrix} u_1 \\ u_2 \\ \vdots \\ u_n \end{bmatrix}, \qquad \mathbf{v} = \begin{bmatrix} v_1 \\ v_2 \\ \vdots \\ v_n \end{bmatrix}$$

where $\{u_1, u_2, \ldots, u_n, v_1, v_2, \ldots, v_n\} \in \mathbb{C}$, their outer product is a matrix:

$$A = \mathbf{u} \otimes \mathbf{v} = \mathbf{u}(\overline{\mathbf{v}})^{\mathrm{T}}$$

$$= \begin{bmatrix} u_1 \\ u_2 \\ \vdots \\ u_n \end{bmatrix} [\overline{v}_1, \overline{v}_2, \cdots, \overline{v}_n]$$

$$= \begin{bmatrix} u_1\overline{v}_1 & u_1\overline{v}_2 & \cdots & u_1\overline{v}_n \\ u_2\overline{v}_1 & u_2\overline{v}_2 & \cdots & u_2\overline{v}_n \\ \vdots & \vdots & \ddots & \vdots \\ u_n\overline{v}_1 & u_n\overline{v}_2 & \cdots & u_n\overline{v}_n \end{bmatrix}.$$

For example,

$$\mathbf{u} = \begin{bmatrix} 2+3i \\ 3+4i \\ 4+5i \end{bmatrix}, \qquad \mathbf{v} = \begin{bmatrix} 5+6i \\ 6+7i \\ 7+8i \end{bmatrix}$$

$$\mathbf{u} \otimes \mathbf{v} = \mathbf{u}(\overline{\mathbf{v}})^{\mathrm{T}} = \begin{bmatrix} 2+3i \\ 3+4i \\ 4+5i \end{bmatrix} [5-6i \quad 6-7i \quad 7-8i]$$

$$= \begin{bmatrix} (2+3i)(5-6i) & (2+3i)(6-7i) & (2+3i)(7-8i) \\ (3+4i)(5-6i) & (3+4i)(6-7i) & (3+4i)(7-8i) \\ (4+5i)(5-6i) & (4+5i)(6-7i) & (4+5i)(7-8i) \end{bmatrix}$$

$$= \begin{bmatrix} 28+3i & 33+4i & 42+5i \\ 39+2i & 46+3i & 53+4i \\ 50+i & 58+2i & 68+3i \end{bmatrix}.$$

3.7 Summary

This chapter should provide the reader with a good background in matrix algebra, especially when involving complex numbers. Hopefully, the above examples, and the following section on Worked Examples, will contextualise the mathematical definitions.

3.7.1 Summary of Formulae

Symmetric Matrix
The symmetric part of a matrix A is given by

$$S = \tfrac{1}{2}\left(A + A^{\mathrm{T}}\right).$$

Anti-Symmetric Matrix
The anti-symmetric part of a matrix A is given by

$$Q = \tfrac{1}{2}\left(A - A^{\mathrm{T}}\right).$$

Conjugate Matrix
Given a matrix A, its conjugate matrix $\overline{A}$ has the signs of its imaginary elements reversed.

Normal Matrix
A complex square matrix A is normal when $\overline{A^{\mathrm{T}}}A = A\overline{A^{\mathrm{T}}}$.

Eigenvector and Eigenvalue

Given a complex matrix A and vector $\mathbf{v}$, then $\mathbf{v}$ is an eigenvector and λ an eigenvalue when

$$A\mathbf{v} = \lambda\mathbf{v}$$
$$A\overline{\mathbf{v}} = \overline{\lambda}\overline{\mathbf{v}}$$
$$|A - \lambda\mathbf{I}| = 0.$$

Conjugate Transpose

Given an $n \times m$ matrix, its conjugate transpose or Hermitian transpose is $A^* = \overline{A^{\mathrm{T}}}$.

Hermitian Matrix

A Hermitian matrix is a complex, square matrix A where $A = \overline{A^{\mathrm{T}}}$.

Orthogonal Matrix

The transpose of an orthogonal matrix is also its inverse: $A^{\mathrm{T}} = A^{-1}$.

Unitary Matrix

A unitary matrix is a complex square matrix whose conjugate transpose is also its inverse: $\overline{A^{\mathrm{T}}}A = I$.

Complex Number as a Matrix

$$1 \equiv \begin{bmatrix} 1 & 0 \\ 0 & 1 \end{bmatrix}, \qquad i \equiv \begin{bmatrix} 0 & -1 \\ 1 & 0 \end{bmatrix}, \qquad a + bi \equiv \begin{bmatrix} a & -b \\ b & a \end{bmatrix}$$

$$i^{-1} \equiv \begin{bmatrix} 0 & 1 \\ -1 & 0 \end{bmatrix}, \qquad e^{i\theta} \equiv \begin{bmatrix} \cos\theta & -\sin\theta \\ \sin\theta & \cos\theta \end{bmatrix}, \qquad |z|^2 = \det\begin{bmatrix} a & -b \\ b & a \end{bmatrix}.$$

Inner Product in $\mathbb{R}^n$

Given two column vectors $\mathbf{u}$ and $\mathbf{v}$ their scalar inner product is

$$\mathbf{u} \cdot \mathbf{v} = \mathbf{u}^{\mathsf{T}}\mathbf{v} = [u_1, u_2, \dots, u_n] \begin{bmatrix} v_1 \\ v_2 \\ \vdots \\ v_n \end{bmatrix} = \sum_{i=1}^{n} u_i v_i.$$

Inner Product in $\mathbb{C}^n$

Given two column vectors $\mathbf{u}$ and $\mathbf{v}$ their complex inner product is

$$\mathbf{u} \cdot \mathbf{v} = \mathbf{u}^{\mathsf{T}}\overline{\mathbf{v}} = [u_1, u_2, \dots, u_n] \begin{bmatrix} \overline{v}_1 \\ \overline{v}_2 \\ \vdots \\ \overline{v}_n \end{bmatrix} = \sum_{i=1}^{n} u_i \overline{v}_i.$$

Distance Between Two Real Vectors

The distance between two vectors $\mathbf{u}$ and $\mathbf{v}$ is $d(\mathbf{u}, \mathbf{v})$:

$$d(\mathbf{u}, \mathbf{v}) = |\mathbf{u} - \mathbf{v}| = \sqrt{(\mathbf{u} - \mathbf{v}) \cdot (\mathbf{u} - \mathbf{v})} = \sqrt{\sum_{i=1}^{n} (u_i - v_i)^2}$$

Distance Between Two Complex Vectors

The distance between two complex vectors $\mathbf{u}$ and $\mathbf{v}$ is $d(\mathbf{u}, \mathbf{v})$:

$$d(\mathbf{u}, \mathbf{v}) = |\mathbf{u} - \mathbf{v}| = \sqrt{(\mathbf{u} - \mathbf{v}) \cdot (\mathbf{u} - \mathbf{v})} = \sqrt{\sum_{i=1}^{n} |u_i - v_i|^2}$$

Outer Product in $\mathbb{R}^n$

Given two real column vectors $\mathbf{u}$ and $\mathbf{v}$, their outer product is

$$A = \mathbf{u} \otimes \mathbf{v} = \mathbf{u}\mathbf{v}^{\mathsf{T}}$$

$$= \begin{bmatrix} u_1 \\ u_2 \\ \vdots \\ u_n \end{bmatrix} [v_1, v_2, \dots, v_n]$$

$$= \begin{bmatrix} u_1 v_1 & u_1 v_2 & \cdots & u_1 v_n \\ u_2 v_1 & u_2 v_2 & \cdots & u_2 v_n \\ \vdots & \vdots & \ddots & \vdots \\ u_n v_1 & u_n v_2 & \cdots & u_n v_n \end{bmatrix}.$$

Outer Product in $\mathbb{C}^n$

Given two complex column vectors **u** and **v**, their outer product is

$$A = \mathbf{u} \otimes \mathbf{v} = \mathbf{u}(\bar{\mathbf{v}})^{\mathrm{T}}$$

$$= \begin{bmatrix} u_1 \\ u_2 \\ \vdots \\ u_n \end{bmatrix} [\bar{v}_1, \bar{v}_2, \ldots, \bar{v}_n]$$

$$= \begin{bmatrix} u_1\bar{v}_1 & u_1\bar{v}_2 & \cdots & u_1\bar{v}_n \\ u_2\bar{v}_1 & u_2\bar{v}_2 & \cdots & u_2\bar{v}_n \\ \vdots & \vdots & \ddots & \vdots \\ u_n\bar{v}_1 & u_n\bar{v}_2 & \cdots & u_n\bar{v}_n \end{bmatrix}.$$

3.8   Worked Examples

3.8.1   Matrix Scaling

Scale matrix A by $-10i$.

$$A = \begin{bmatrix} -9i & 100i \\ 1+i & 4i \end{bmatrix}.$$

Solution: Multiply all the elements of A by $-10i$.

$$-10i\,A = \begin{bmatrix} -90 & 1000 \\ 10-10i & 40 \end{bmatrix}.$$

3.8.2   Common Factor

Identify the common factor of matrix A.

$$A = \begin{bmatrix} 4i & -12i \\ 32 & 40i \end{bmatrix}.$$

Solution: $4i$ is the common factor of the elements of A.

$$A = 4i \begin{bmatrix} 1 & -3 \\ -8i & 10 \end{bmatrix}.$$

3.8.3 Matrix Multiplication

Compute the product AB.

$$A = \begin{bmatrix} 2+i & 2-i \\ 5 & 4+5i \end{bmatrix}, \quad B = \begin{bmatrix} 3+i & 4-i \\ 2+3i & 5+2i \end{bmatrix}.$$

Solution: Expand the product AB according to the rules of matrix algebra.

$$\begin{aligned} AB &= \begin{bmatrix} 2+i & 2-i \\ 5 & 4+5i \end{bmatrix} \begin{bmatrix} 3+i & 4-i \\ 2+3i & 5+2i \end{bmatrix} \\ &= \begin{bmatrix} (2+i)(3+i)+(2-i)(2+3i) & (2+i)(4-i)+(2-i)(5+2i) \\ 5(3+i)+(4+5i)(2+3i) & 5(4-i)+(4+5i)(5+2i) \end{bmatrix} \\ &= \begin{bmatrix} (5+5i)+(7+4i) & (9+2i)+(12-i) \\ (15+5i)+(-7+22i) & (20-5i)+(10+33i) \end{bmatrix} \\ &= \begin{bmatrix} 12+9i & 21+i \\ 8+27i & 30+28i \end{bmatrix}. \end{aligned}$$

3.8.4 Determinant of a Matrix

Find the determinant of A.

$$A = \begin{bmatrix} i & 2+2i & 3+3i \\ 1+2i & 2+3i & 3+4i \\ 1+3i & 2+4i & 3+5i \end{bmatrix}.$$

Solution: Use Sarrus's rule.

$$\begin{aligned} \det[A] &= i(2+3i)(3+5i)+(2+2i)(3+4i)(1+3i)+(3+3i)(1+2i)(2+4i) \\ &\quad -(1+i)(3+4i)(2+4i)-(2+2i)(1+2i)(3+5i)-(3+3i)(2+3i)(1+3i) \\ &= i(-9+19i)+(2+2i)(-9+13i)+(3+3i)(-6+8i) \\ &\quad -i(-10+20i)-(2+2i)(-7+11i)-(3+3i)(-7+9i) \\ &= (-19-9i)+(-44+8i)+(-42+6i)-(-20-10i)-(-36+8i)-(-48+6i) \\ &= -1+i. \end{aligned}$$

3.8.5 Transpose Matrix

State the transpose of matrix A.

$$A = \begin{bmatrix} i & 2+2i & 3+3i \\ 1+2i & 2+3i & 3+4i \\ 1+3i & 2+4i & 3+5i \end{bmatrix}.$$

Solution: Exchange A's rows and columns.

$$A^\mathsf{T} = \begin{bmatrix} i & 1+2i & 1+3i \\ 2+2i & 2+3i & 2+4i \\ 3+3i & 3+4i & 3+5i \end{bmatrix}.$$

3.8.6   Symmetric Matrix

Compute the symmetric part of A.

$$A = \begin{bmatrix} i & 2+2i & 3+3i \\ 1+2i & 2+3i & 3+4i \\ 1+3i & 2+4i & 3+5i \end{bmatrix}.$$

Solution: Use S to compute the symmetric part of A.

$$S = \begin{bmatrix} a_{11} & \dfrac{s_3}{2} & \dfrac{s_2}{2} \\ \dfrac{s_3}{2} & a_{22} & \dfrac{s_1}{2} \\ \dfrac{s_2}{2} & \dfrac{s_1}{2} & a_{33} \end{bmatrix}$$

where

$$s_1 = a_{23} + a_{32} = 3 + 4i + 2 + 4i = 5 + 8i$$
$$s_2 = a_{13} + a_{31} = 3 + 3i + 1 + 3i = 4 + 6i$$
$$s_3 = a_{12} + a_{21} = 2 + 2i + 1 + 2i = 3 + 4i$$

$$S = \begin{bmatrix} i & 1.5+2i & 2+3i \\ 1.5+2i & 2+3i & 2.5+4i \\ 2+3i & 2.5+4i & 3+5i \end{bmatrix}.$$

3.8.7   Anti-symmetric Matrix

Compute the anti-symmetric part of A and show that together with the symmetric part, creates A.

$$A = \begin{bmatrix} i & 2+2i & 3+3i \\ 1+2i & 2+3i & 3+4i \\ 1+3i & 2+4i & 3+5i \end{bmatrix}.$$

Solution: Use Q to compute the anti-symmetric part of A.

$$Q = \begin{bmatrix} 0 & \dfrac{q_3}{2} & -\dfrac{q_2}{2} \\ -\dfrac{q_3}{2} & 0 & \dfrac{q_1}{2} \\ \dfrac{q_2}{2} & -\dfrac{q_1}{2} & 0 \end{bmatrix}$$

where

$$q_1 = a_{23} - a_{32} = (3+4i) - (2+4i) = 1$$
$$q_2 = a_{31} - a_{13} = (1+3i) - (3+3i) = -2$$
$$q_3 = a_{12} - a_{21} = (2+2i) - (1+2i) = 1$$

$$Q = \begin{bmatrix} 0 & 0.5 & 1 \\ -0.5 & 0 & 0.5 \\ -1 & -0.5 & 0 \end{bmatrix}.$$

$$S + Q = \begin{bmatrix} i & 1.5+2i & 2+3i \\ 1.5+2i & 2+3i & 2.5+4i \\ 2+3i & 2.5+4i & 3+5i \end{bmatrix} + \begin{bmatrix} 0 & 0.5 & 1 \\ -0.5 & 0 & 0.5 \\ -1 & -0.5 & 0 \end{bmatrix}$$

$$A = \begin{bmatrix} i & 2+2i & 3+3i \\ 1+2i & 2+3i & 3+4i \\ 1+3i & 2+4i & 3+5i \end{bmatrix}.$$

3.8.8 Cofactor Matrix

Determine the cofactor matrix for A.

$$A = \begin{bmatrix} i & 2+2i & 3+3i \\ 1+2i & 2+3i & 3+4i \\ 1+3i & 2+4i & 3+5i \end{bmatrix}.$$

Solution: Use the following cofactor template.

$$\text{cofactor matrix of } A = \begin{bmatrix} A_{11} & A_{12} & A_{13} \\ A_{21} & A_{22} & A_{23} \\ A_{31} & A_{32} & A_{33} \end{bmatrix}$$

where

$$A_{11} = + \begin{vmatrix} a_{22} & a_{23} \\ a_{32} & a_{33} \end{vmatrix} = + \begin{vmatrix} 2+3i & 3+4i \\ 2+4i & 3+5i \end{vmatrix} = 1 - i$$

$$A_{12} = - \begin{vmatrix} a_{21} & a_{23} \\ a_{31} & a_{33} \end{vmatrix} = - \begin{vmatrix} 1+2i & 3+4i \\ 1+3i & 3+5i \end{vmatrix} = -2 + 2i$$

$$A_{13} = + \begin{vmatrix} a_{21} & a_{22} \\ a_{31} & a_{32} \end{vmatrix} = + \begin{vmatrix} 1+2i & 2+3i \\ 1+3i & 2+4i \end{vmatrix} = 1 - i$$

$$A_{21} = - \begin{vmatrix} a_{12} & a_{13} \\ a_{32} & a_{33} \end{vmatrix} = - \begin{vmatrix} 2+2i & 3+3i \\ 2+4i & 3+5i \end{vmatrix} = -2 + 2i$$

$$A_{22} = + \begin{vmatrix} a_{11} & a_{13} \\ a_{31} & a_{33} \end{vmatrix} = + \begin{vmatrix} i & 3+3i \\ 1+3i & 3+5i \end{vmatrix} = 1 - 9i$$

$$A_{23} = - \begin{vmatrix} a_{11} & a_{12} \\ a_{31} & a_{32} \end{vmatrix} = - \begin{vmatrix} i & 2+2i \\ 1+3i & 2+4i \end{vmatrix} = 6i$$

$$A_{31} = + \begin{vmatrix} a_{12} & a_{13} \\ a_{22} & a_{23} \end{vmatrix} = + \begin{vmatrix} 2+2i & 3+3i \\ 2+3i & 3+4i \end{vmatrix} = 1 - i$$

$$A_{32} = - \begin{vmatrix} a_{11} & a_{13} \\ a_{21} & a_{23} \end{vmatrix} = - \begin{vmatrix} i & 3+3i \\ 1+2i & 3+4i \end{vmatrix} = 1 + 6i$$

$$A_{33} = + \begin{vmatrix} a_{11} & a_{12} \\ a_{21} & a_{22} \end{vmatrix} = + \begin{vmatrix} i & 2+2i \\ 1+2i & 2+3i \end{vmatrix} = -1 - 4i$$

therefore, the cofactor matrix of A is

$$\begin{bmatrix} 1-i & -2+2i & 1-i \\ -2+2i & 1-9i & 6i \\ 1-i & 1+6i & -1-4i \end{bmatrix}.$$

3.8.9   Inverse Matrix

Find the inverse of matrix A using the determinant and cofactor matrix from above.

$$A = \begin{bmatrix} i & 2+2i & 3+3i \\ 1+2i & 2+3i & 3+4i \\ 1+3i & 2+4i & 3+5i \end{bmatrix}.$$

Solution: Invert matrix A using the cofactor matrix algorithm.

$$(\text{cofactor matrix of } A) = \begin{bmatrix} 1-i & -2+2i & 1-i \\ -2+2i & 1-9i & 6i \\ 1-i & 1+6i & -1-4i \end{bmatrix}$$

$$(\text{cofactor matrix of } A)^{\text{T}} = \begin{bmatrix} 1-i & -2+2i & 1-i \\ -2+2i & 1-9i & 1+6i \\ 1-i & 6i & -1-4i \end{bmatrix}$$

$$|A| = -1+i$$

$$A^{-1} = \frac{1}{-1+i} \begin{bmatrix} 1-i & -2+2i & 1-i \\ -2+2i & 1-9i & 1+6i \\ 1-i & 6i & -1-4i \end{bmatrix}$$

$$= \frac{-1-i}{2} \begin{bmatrix} 1-i & -2+2i & 1-i \\ -2+2i & 1-9i & 1+6i \\ 1-i & 6i & -1-4i \end{bmatrix}$$

$$= \begin{bmatrix} -1 & 2 & -1 \\ 2 & -5+4i & 2.5-3.5i \\ -1 & 3-3i & -1.5+2.5i \end{bmatrix}.$$

Let's check this result by multiplying A by A^{-1} which must equal I:

$$AA^{-1} = \begin{bmatrix} i & 2+2i & 3+3i \\ 1+2i & 2+3i & 3+4i \\ 1+3i & 2+4i & 3+5i \end{bmatrix} \begin{bmatrix} -1 & 2 & -1 \\ 2 & -5+4i & 2.5-3.5i \\ -1 & 3-3i & -1.5+2.5i \end{bmatrix}$$

$a_{11} = -i + 2(2+2i) - (3+3i) = 1$

$a_{12} = 2i + (2+2i)(-5+4i) + (3+3i)(3-3i) = 0$

$a_{13} = -i + (2+2i)(2.5-3.5i) + (3+3i)(-1.5+2.5i) = 0$

$a_{21} = -(1+2i) + 2(2+3i) - (3+4i) = 0$

$a_{22} = 2(1+2i) + (2+3i)(-5+4i) + (3+4i)(3-3i) = 1$

$a_{23} = -(1+2i) + (2+3i)(2.5-3.5i) + (3+4i)(-1.5+2.5i) = 0$

$a_{31} = -(1+3i) + 2(2+4i) - (3+5i) = 0$

$a_{32} = 2(1+3i) + (2+4i)(-5+4i) + (3+5i)(3-3i) = 0$

$a_{33} = -(1+3i) + (2+4i)(2.5-3.5i) + (3+5i)(-1.5+2.5i) = 1$

$$= \begin{bmatrix} 1 & 0 & 0 \\ 0 & 1 & 0 \\ 0 & 0 & 1 \end{bmatrix}.$$

3.8.10   Conjugate Matrix

State the conjugate matrix of A.

$$A = \begin{bmatrix} 1-2i & 2 & 3+2i \\ -3i & i & 2-4i \\ 3 & -2i & 4+2i \end{bmatrix}.$$

Solution: Conjugate the individual elements of A.

$$\overline{A} = \begin{bmatrix} 1+2i & 2 & 3-2i \\ 3i & -i & 2+4i \\ 3 & 2i & 4-2i \end{bmatrix}.$$

3.8.11   Complex Eigenvectors and Eigenvalues

Find the eigenvectors and eigenvalues for matrix A.

$$A = \begin{bmatrix} 6 & -1 \\ 17 & -2 \end{bmatrix}.$$

Solution: Solve the characteristic equation to determine the eigenvectors and eigenvalues. The characteristic equation is

$$\begin{vmatrix} 6-\lambda & -1 \\ 17 & -2-\lambda \end{vmatrix} = (6-\lambda)(-2-\lambda) - (-1)17 = 0$$

$$0 = -12 - 6\lambda + 2\lambda + \lambda^2 + 17$$

$$= \lambda^2 - 4\lambda + 5$$

which has roots given by

$$\lambda = \frac{4 \pm \sqrt{16-20}}{2} = 2 \pm i$$

and form a complex conjugate pair.

$$Av = (2+i)v$$

$$\begin{bmatrix} 6 & -1 \\ 17 & -2 \end{bmatrix} \begin{bmatrix} x \\ y \end{bmatrix} = (2+i) \begin{bmatrix} 1 & 0 \\ 0 & 1 \end{bmatrix} \begin{bmatrix} x \\ y \end{bmatrix}$$

$$\begin{bmatrix} 4-i & -1 \\ 17 & -(4+i) \end{bmatrix} \begin{bmatrix} x \\ y \end{bmatrix} = \begin{bmatrix} 0 \\ 0 \end{bmatrix}$$

which provides two equations:

$$(4-i)x - y = 0$$
$$17x - (4+i)y = 0.$$

However, the determinant of the associated matrix is zero:

$$\begin{vmatrix} 4-i & -1 \\ 17 & -4-i \end{vmatrix} = 0$$

and either equation can be used:

$$y = (4-i)x. \tag{3.13}$$

To simplify (3.13) let $k = x$:

$$v = \begin{bmatrix} x \\ y \end{bmatrix} = \begin{bmatrix} k \\ k(4-i) \end{bmatrix} = k \begin{bmatrix} 1 \\ 4-i \end{bmatrix}$$

where k is an arbitrary scalar.

The second eigenvalue $2 - i$, has a corresponding eigenvector:

$$v = k \begin{bmatrix} 1 \\ 4+i \end{bmatrix}.$$

The eigenvalues and eigenvectors are

$$\lambda = 2+i, \quad v = \begin{bmatrix} 1 \\ 4-i \end{bmatrix}$$

$$\lambda = 2-i, \quad v = \begin{bmatrix} 1 \\ 4+i \end{bmatrix}$$

now let's show that $Av = \lambda v$ for both pairs:

$$Av = \begin{bmatrix} 6 & -1 \\ 17 & -2 \end{bmatrix} \begin{bmatrix} 1 \\ 4-i \end{bmatrix} = \begin{bmatrix} 6-4+i \\ 17-8+2i \end{bmatrix} = \begin{bmatrix} 2+i \\ 9+2i \end{bmatrix}$$

$$\lambda v = (2+i) \begin{bmatrix} 1 \\ 4-i \end{bmatrix} = \begin{bmatrix} 2+i \\ (2+i)(4-i) \end{bmatrix} = \begin{bmatrix} 2+i \\ 9+2i \end{bmatrix},$$

$$Av = \begin{bmatrix} 6 & -1 \\ 17 & -2 \end{bmatrix} \begin{bmatrix} 1 \\ 4+i \end{bmatrix} = \begin{bmatrix} 6-4-i \\ 17-8-2i \end{bmatrix} = \begin{bmatrix} 2-i \\ 9-2i \end{bmatrix}$$

$$\lambda v = (2-i) \begin{bmatrix} 1 \\ 4+i \end{bmatrix} = \begin{bmatrix} 2-i \\ (2-i)(4+i) \end{bmatrix} = \begin{bmatrix} 2-i \\ 9-2i \end{bmatrix}.$$

3.8.12 Conjugate Transpose Matrix

Find the conjugate transpose of A.

$$A = \begin{bmatrix} 2-3i & -i & 4 \\ 4+2i & 6i & 0 \\ 2+5i & 2+2i & 3+3i \\ 1 & 3i & 9+9i \end{bmatrix}.$$

Solution: Conjugate the individual elements of A and transpose the rows and columns.

$$\overline{A} = \begin{bmatrix} 2+3i & i & 4 \\ 4-2i & -6i & 0 \\ 2-5i & 2-2i & 3-3i \\ 1 & -3i & 9-9i \end{bmatrix}$$

$$\overline{A}^T = \begin{bmatrix} 2+3i & 4-2i & 2-5i & 1 \\ i & -6i & 2-2i & -3i \\ 4 & 0 & 3-3i & 9-9i \end{bmatrix}.$$

3.8.13 Hermitian Matrix

Show that A is a Hermitian matrix and B is not.

$$A = \begin{bmatrix} 1 & 1+4i & 5i \\ 1-4i & 2 & 8+3i \\ -5i & 8-3i & 0 \end{bmatrix}, \quad B = \begin{bmatrix} 1 & 1+4i & 5i \\ 1-4i & 2 & 8+3i \\ -5i & 8-3i & 2i \end{bmatrix}.$$

Solution: Show that $\overline{A}^T = A$ and $\overline{B}^T \neq B$.

Find $\overline{A}$

$$\overline{A} = \begin{bmatrix} 1 & 1-4i & -5i \\ 1+4i & 2 & 8-3i \\ 5i & 8+3i & 0 \end{bmatrix}$$

whose transpose $\overline{A}^T$ is

$$\overline{A}^T = \begin{bmatrix} 1 & 1+4i & 5i \\ 1-4i & 2 & 8+3i \\ -5i & 8-3i & 0 \end{bmatrix} = A = A^H.$$

Find $\overline{B}$

$$\overline{B} = \begin{bmatrix} 1 & 1-4i & -5i \\ 1+4i & 2 & 8-3i \\ 5i & 8+3i & -2i \end{bmatrix}$$

whose transpose $\overline{B}^T$ is

$$\overline{B}^T = \begin{bmatrix} 1 & 1+4i & 5i \\ 1-4i & 2 & 8+3i \\ -5i & 8-3i & -2i \end{bmatrix} \neq B \neq B^H.$$

3.8.14 Orthogonal Matrix

Show that A is orthogonal.

$$A = \begin{bmatrix} i & \sqrt{2} & 0 \\ -\sqrt{2} & i & 0 \\ 0 & 0 & -i \end{bmatrix}.$$

Solution: Show that $AA^T = I$.

$$AA^T = \begin{bmatrix} i & \sqrt{2} & 0 \\ -\sqrt{2} & i & 0 \\ 0 & 0 & -i \end{bmatrix} \begin{bmatrix} i & -\sqrt{2} & 0 \\ \sqrt{2} & i & 0 \\ 0 & 0 & -i \end{bmatrix}$$

$$= \begin{bmatrix} 1 & 0 & 0 \\ 0 & 1 & 0 \\ 0 & 0 & 1 \end{bmatrix} = I.$$

Therefore, A is orthogonal.

3.8.15   Unitary Matrix

Show that A is a unitary matrix.

$$A = \begin{bmatrix} \frac{1}{\sqrt{2}} & \frac{1}{\sqrt{2}} & 0 \\ -\frac{i}{\sqrt{2}} & \frac{i}{\sqrt{2}} & 0 \\ 0 & 0 & -i \end{bmatrix}$$

Solution: Show that $\overline{A^{\mathrm{T}}} A = A \overline{A^{\mathrm{T}}} = I$.

$$A^{\mathrm{T}} = \begin{bmatrix} \frac{1}{\sqrt{2}} & -\frac{i}{\sqrt{2}} & 0 \\ \frac{1}{\sqrt{2}} & \frac{i}{\sqrt{2}} & 0 \\ 0 & 0 & -i \end{bmatrix}$$

$$\overline{A^{\mathrm{T}}} = \begin{bmatrix} \frac{1}{\sqrt{2}} & \frac{i}{\sqrt{2}} & 0 \\ \frac{1}{\sqrt{2}} & -\frac{i}{\sqrt{2}} & 0 \\ 0 & 0 & i \end{bmatrix}$$

$$\overline{A^{\mathrm{T}}} A = \begin{bmatrix} \frac{1}{\sqrt{2}} & \frac{i}{\sqrt{2}} & 0 \\ \frac{1}{\sqrt{2}} & -\frac{i}{\sqrt{2}} & 0 \\ 0 & 0 & -i \end{bmatrix} \begin{bmatrix} \frac{1}{\sqrt{2}} & \frac{1}{\sqrt{2}} & 0 \\ -\frac{i}{\sqrt{2}} & \frac{i}{\sqrt{2}} & 0 \\ 0 & 0 & -i \end{bmatrix} = \begin{bmatrix} 1 & 0 & 0 \\ 0 & 1 & 0 \\ 0 & 0 & 1 \end{bmatrix}$$

$$A \overline{A^{\mathrm{T}}} = \begin{bmatrix} \frac{1}{\sqrt{2}} & \frac{1}{\sqrt{2}} & 0 \\ -\frac{i}{\sqrt{2}} & \frac{i}{\sqrt{2}} & 0 \\ 0 & 0 & -i \end{bmatrix} \begin{bmatrix} \frac{1}{\sqrt{2}} & \frac{i}{\sqrt{2}} & 0 \\ \frac{1}{\sqrt{2}} & -\frac{i}{\sqrt{2}} & 0 \\ 0 & 0 & -i \end{bmatrix} = \begin{bmatrix} 1 & 0 & 0 \\ 0 & 1 & 0 \\ 0 & 0 & 1 \end{bmatrix}$$

therefore, A is a unitary matrix.

3.8.16   Complex Vector Addition

Given $\mathbf{u}$, $\mathbf{v}$ and $\mathbf{w}$, compute $\mathbf{u} + \mathbf{v}$, $\mathbf{u} + \mathbf{w}$ and $\mathbf{v} + \mathbf{w}$.

$$\mathbf{u} = \begin{bmatrix} 2 - 3i \\ 1 + 4i \\ 3 + 5i \\ i \end{bmatrix}, \qquad \mathbf{v} = \begin{bmatrix} 4 - 4i \\ 6 + 4i \\ 5 + 3i \\ 3 \end{bmatrix}, \qquad \mathbf{w} = \begin{bmatrix} 2 \\ 6 - 4i \\ -5 - 3i \\ -i \end{bmatrix}.$$

Solution: Add the respective elements.

$$\mathbf{u} + \mathbf{v} = \begin{bmatrix} 6 - 7i \\ 7 + 8i \\ 8 + 8i \\ 3 + i \end{bmatrix}, \quad \mathbf{u} + \mathbf{w} = \begin{bmatrix} 4 - 3i \\ 7 \\ -2 + 2i \\ 0 \end{bmatrix}, \quad \mathbf{v} + \mathbf{w} = \begin{bmatrix} 6 - 4i \\ 12 \\ 0 \\ 3 - i \end{bmatrix}.$$

3.8.17 Complex Inner Product

Find the inner product of $\mathbf{u}$ and $\mathbf{v}$.

$$\mathbf{u} = \begin{bmatrix} 2 + 2i \\ 3 - 3i \\ 5 + 3i \end{bmatrix}, \quad \mathbf{v} = \begin{bmatrix} 3 - i \\ 4 + 3i \\ 1 + 2i \end{bmatrix}.$$

Solution: Multiply $\mathbf{u}$ and $\mathbf{v}$ using $\mathbf{u}^T\bar{\mathbf{v}}$.

$$
\begin{aligned}
\mathbf{u} \cdot \mathbf{v} = \mathbf{u}^T\bar{\mathbf{v}} &= [2 + 2i, \ 3 - 3i, \ 5 + 3i] \begin{bmatrix} 3 + i \\ 4 - 3i \\ 1 - 2i \end{bmatrix} \\
&= (2 + 2i)(3 + i) + (3 - 3i)(4 - 3i) + (5 + 3i)(1 - 2i) \\
&= 4 + 8i + 3 - 21i + 11 - 7i \\
&= 18 - 20i.
\end{aligned}
$$

3.8.18 Complex Norm

Find the complex norms of $\mathbf{u}$ and $\mathbf{v}$.

$$\mathbf{u} = \begin{bmatrix} 2 + 2i \\ 3 - 3i \\ 5 + 3i \end{bmatrix}, \quad \mathbf{v} = \begin{bmatrix} 3 - i \\ 4 + 3i \\ 1 + 2i \end{bmatrix}.$$

Solution: For each vector, take the square-root of the sum of the squares of the norms.

$$
\begin{aligned}
\|\mathbf{u}\| &= \sqrt{|2 + 2i|^2 + |3 - 3i|^2 + |5 + 3i|^2} \\
&= \sqrt{(2^2 + 2^2) + (3^2 + 3^2) + (5^2 + 3^2)} \\
&= \sqrt{8 + 18 + 34} = \sqrt{60}.
\end{aligned}
$$

$$\|\mathbf{v}\| = \sqrt{|3 - i|^2 + |4 + 3i|^2 + |1 + 2i|^2}$$
$$= \sqrt{(3^2 + 1^2) + (4^2 + 3^2) + (1^2 + 2^2)}$$
$$= \sqrt{10 + 25 + 5} = \sqrt{40}.$$

3.8.19   Distance Between Complex Vectors

Find the distance between $\mathbf{u}$ and $\mathbf{v}$.

$$\mathbf{u} = \begin{bmatrix} 2 + 2i \\ 3 - 3i \\ 5 + 3i \end{bmatrix}, \quad \mathbf{v} = \begin{bmatrix} 3 - i \\ 4 + 3i \\ 1 + 2i \end{bmatrix}.$$

Solution: Use $d(\mathbf{u}, \mathbf{v}) = |\mathbf{u} - \mathbf{v}|$.

$$d(\mathbf{u}, \mathbf{v}) = |\mathbf{u} - \mathbf{v}|$$
$$= |(-1 + 3i), (-1 - 6i), (3 + i)|$$
$$= \sqrt{1^2 + 3^2 + 1^2 + 6^2 + 3^2 + 1^2}$$
$$= \sqrt{1 + 9 + 1 + 36 + 9 + 1} = \sqrt{57}.$$

3.8.20   Complex Outer Product

Compute $\mathbf{u} \otimes \mathbf{v}$.

$$\mathbf{u} = \begin{bmatrix} 1 + i \\ 2 + 2i \\ 3 + 3i \end{bmatrix}, \quad \mathbf{v} = \begin{bmatrix} 1 - i \\ 2 - 2i \\ 3 - 3i \end{bmatrix}.$$

Solution: Use $\mathbf{u} \otimes \mathbf{v} = \mathbf{u}(\bar{\mathbf{v}})^{\mathrm{T}}$.

$$\mathbf{u} \otimes \mathbf{v} = \mathbf{u}(\overline{\mathbf{v}})^{\mathrm{T}} = \begin{bmatrix} 1+i \\ 2+2i \\ 3+3i \end{bmatrix} [1-i \quad 2-2i \quad 3-3i]$$

$$= \begin{bmatrix} (1+i)(1-1i) & (1+i)(2-2i) & (1+i)(3-3i) \\ (2+2i)(1-i) & (2+2i)(2-2i) & (2+2i)(3-3i) \\ (3+3i)(1-i) & (3+3i)(2-2i) & (3+3i)(3-3i) \end{bmatrix}$$

$$= \begin{bmatrix} 2 & 4 & 6 \\ 4 & 8 & 12 \\ 6 & 12 & 18 \end{bmatrix}.$$

Reference

1. Vince JA (2017) Mathematics for computer graphics, 5th edn. Springer, Berlin

Chapter 4
Quaternions

4.1 Introduction

In this chapter I describe the invention of quaternions by Sir William Rowan Hamilton, and their associated complex algebra. However, as so often happens in mathematics, someone-else had already touched upon the subject before Hamilton, as we shall see. If you are interested in the historical development of quaternions, vectors and geometric algebra, then you must read Michael Crowe's book *A History of Vector Analysis* [1].

Quaternions are a natural extension to complex numbers and have a variety of forms and operations such as pure, unit, binary, inverse, etc., all of which are described. The chapter contains many worked examples, and leads the reader towards the point where they can be used to rotate vectors in $\mathbb{R}^3$.

4.1.1 History of Quaternions

In Chap. 1 we saw how Wessel and Argand both invented the complex plane, and used it to visualise complex numbers. It was unfortunate for both men that they didn't have access to today's ubiquitous publishing network, and the world-wide web. Nevertheless, priority was – and still is – decided by who gets to the printing presses first. But as we saw with Wessel, even being first into print didn't guarantee fame.

A similar story surrounds the invention of quaternions. Sir William Rowan Hamilton is recognised as the inventor of quaternion algebra, which became the first non-commutative algebra to be discovered. One can imagine the elation he felt when finding a solution to a problem he had been thinking about for over a decade!

The invention provided the first mathematical framework for manipulating vectorial quantities, although this was to be refined by the American theoretical physicist, chemist, and mathematician, Josiah Willard Gibbs (1839–1903). Although Hamil-

© Springer International Publishing AG, part of Springer Nature 2018
J. Vince, *Imaginary Mathematics for Computer Science*,
https://doi.org/10.1007/978-3-319-94637-5_4

ton had arrived at his invention through an algebraic route, it was obvious to him
that quaternions had significant geometric potential, and he immediately started to
explore their vectorial and rotational properties.

Unbeknown to Hamilton – and virtually everyone else at the time – the French
social reformer, and brilliant recreational mathematician Benjamin Olinde Rodrigues
(1795–1851), had already published a paper in 1840 describing how to represent two
successive rotations about different axes, by a single rotation about a third axis [2].
What is more, Rodrigues expressed his solution using a scalar and a 3D axis, which
pre-empted Hamilton's own approach using a scalar and a vector, by three years!
Simon Altmann has probably done more than any other person to set this record
straight, and has published his views widely [3–6]. However, for the moment, let's
continue with Hamilton's imaginary algebra.

The very existence of complex numbers presented a tantalising question for mathematicians of the 18 and 19th centuries. Could there be a 3D equivalent? The answer
to this question was not obvious, and many gifted mathematicians, including Gauss,
Möbius, Grassmann, and Hamilton had been searching for the answer.

Hamilton's research is well documented and covers a period from the early 1830s
to 1843, when he invented quaternions. And for a further 22 years, until his death
in 1865, he was preoccupied with the subject. By 1833 he had shown that complex
numbers form an algebra of couples, i.e. ordered pairs [7].

As a 2D complex number is represented by $a + bi$, Hamilton conjectured that a
3D complex number could be represented by the triple, $a + bi + cj$, where i^2 and
j^2 are imaginary units that equal -1. However, the product of two such triples raises
a problem with their algebraic expansion:

$$z_1 = a_1 + b_1 i + c_1 j$$
$$z_2 = a_2 + b_2 i + c_2 j$$
$$z_1 z_2 = (a_1 + b_1 i + c_1 j)(a_2 + b_2 i + c_2 j)$$
$$= a_1 a_2 + a_1 b_2 i + a_1 c_2 j$$
$$+ b_1 a_2 i + b_1 b_2 i^2 + b_1 c_2 ij + c_1 a_2 j + c_1 b_2 ji + c_1 c_2 j^2$$
$$= (a_1 a_2 - b_1 b_2 - c_1 c_2) + (a_1 b_2 + b_1 a_2)i + (a_1 c_2 + c_1 a_2)j + b_1 c_2 ij + c_1 b_2 ji$$

the operation almost closes – apart from the terms involving ij and ji. Even if we
assume that $ji = -ij$, we are still left with

$$(b_1 c_2 - c_1 b_2)ij.$$

This presented a real problem for Hamilton and he toiled for over a decade trying to
resolve it. Then, on 16 October, 1843, whilst walking with his wife, Lady Hamilton,
along the Royal Canal in Ireland to preside at a meeting of the Royal Irish Academy
[8], a flash of inspiration came to him where he saw the solution as a quadruple,
rather than a triple. Instead of using two imaginary terms, three terms provided the
extra permutations necessary to resolve products like ij.

The solution was $z = a + bi + cj + dk$ where $i^2 = j^2 = k^2 = -1$. And because of the four terms, Hamilton gave the name *quaternion*. Hamilton took the opportunity to record the event in stone, by carving the rules into the wall of Broome bridge, which he was passing at the time. Although his original inscription has not withstood years of Irish weather, a more permanent plaque now replaces it.

When Hamilton invented quaternions, he also created all sorts of names such as *tensor*, *versor* and *vector* to describe their attributes. As the inventor, it was Hamilton's prerogative to choose whatever names he wanted, and at the time, such names were associated with the notation of the period. For example, he called the quaternion's real part a *scalar*, and the imaginary part a *vector*. However today, a vector does not have any imaginary associations, which has slightly confused how quaternions are interpreted.

Let's examine the algebra of quaternions which form the set $\mathbb{H}$ in recognition of Hamilton's achievement.

4.2 Some Algebraic History

Hamilton defined a quaternion q, and its associated rules as

$$q = s + ia + jb + kc, \quad \{s, a, b, c\} \in \mathbb{R}$$

and

$$i^2 = j^2 = k^2 = ijk = -1$$

$$ij = k, \qquad jk = i, \qquad ki = j$$
$$ji = -k, \quad kj = -i, \quad ik = -j$$

References [9–11]; but today, we tend to write a quaternion as

$$q = s + ai + bj + ck.$$

Observe from Hamilton's rules, Table 4.1, how the occurrence of ij is replaced by k, and $ji = -k$. The extra imaginary k term is key to the cyclic patterns $ij = k$, $jk = i$, and $ki = j$, which are very similar to the cross product of two unit Cartesian vectors:

$$\mathbf{i} \times \mathbf{j} = \mathbf{k}, \quad \mathbf{j} \times \mathbf{k} = \mathbf{i}, \quad \mathbf{k} \times \mathbf{i} = \mathbf{j}.$$

In fact, this similarity is no coincidence, as Hamilton also invented the scalar and vector products. However, although quaternions provide an algebraic framework to describe vectors, one must acknowledge that vectorial quantities had been studied for many years prior to Hamilton.

Table 4.1 Multiplication table for quaternion imaginaries

	i	j	k
i	-1	k	$-j$
j	$-k$	-1	i
k	j	$-i$	-1

Hamilton also saw that the i, j, k terms could represent three Cartesian unit vectors **i**, **j** and **k**, which had to possess imaginary qualities. i.e. $\mathbf{i}^2 = -1$, etc., which didn't go down well with some mathematicians and scientists who were suspicious of the need to involve so many imaginary terms.

Hamilton's motivation to search for a 3D equivalent of complex numbers was part algebraic, and part geometric. For as a complex number can be represented by a couple (an ordered pair), and is capable of rotating points on the complex plane, then perhaps a *triple* could rotate points in space? In the end, a triple had to be replaced by a a quadruple – a quaternion.

One can regard Hamilton's rules from two perspectives. The first, is that they are an algebraic consequence of combining three imaginary terms. The second, is that they reflect an underlying geometric structure of space. The latter interpretation was adopted by P. G. Tait, and outlined in his book *An Elementary Treatise on Quaternions*. Tait's approach assumes three unit vectors **i**, **j**, **k** aligned with the x-, y-, z-axis respectively:

> The result of the multiplication of **i** into **j** or **ij** is defined to be the turning of **j** through a right angle in the plane perpendicular to **i** in the positive direction, in other words, the operation of **i** on **j** turns it round so as to make it coincide with **k**; and therefore briefly **ij** = **k**.

> To be consistent it is requisite to admit that if **i** instead of operating on **j** had operated on any other unit vector perpendicular to **i** in the plane yz, it would have turned it through a right-angle in the same direction, so that **ik** can be nothing else than $-\mathbf{j}$.

> Extending to other unit vectors the definition which we have illustrated by referring to **i**, it is evident that **j** operating on **k** must bring it round to **i**, or **jk** = **i** [12].

Tait's explanation is illustrated in Fig. 4.1a–d. Figure 4.1a shows the original alignment of **i**, **j**, **k**. Figure 4.1b shows the effect of turning **j** into **k**. Figure 4.1c shows the turning of **k** into **i**, and Fig. 4.1d shows the turning of **i** in to **j**.

So far, there is no mention of imaginary quantities – we just have

$$\mathbf{ij} = \mathbf{k}, \quad \mathbf{jk} = \mathbf{i}, \quad \mathbf{ki} = \mathbf{j}$$
$$\mathbf{ji} = -\mathbf{k}, \quad \mathbf{kj} = -\mathbf{i}, \quad \mathbf{ik} = -\mathbf{j}.$$

If we assume that these vectors obey the distributive and associative axioms of algebra, their imaginary qualities are exposed. For example:

$$\mathbf{ij} = \mathbf{k}$$

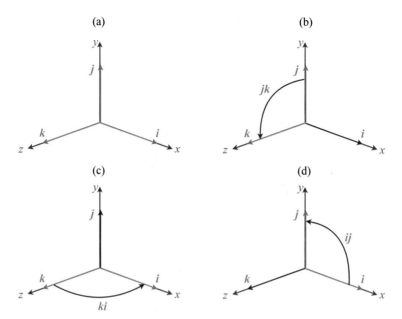

Fig. 4.1 Interpreting the products **ij**, **jk**, **ki**

and multiplying throughout by **i**:

$$\mathbf{iij = ik = -j}$$

therefore,

$$\mathbf{ii = i^2 = -1}.$$

Similarly, we can show that $\mathbf{j^2 = k^2 = -1}$.
    Next:

$$\mathbf{ijk = i(jk) = ii = i^2 = -1}.$$

Thus, simply by declaring the action of the cross-product, Hamilton's rules emerge, with all of their imaginary features. Tait also made the following observation:

A very curious speculation, due to Servois, and published in 1813 in Gergonne's Annales is the only one, so far has been discovered, in which the slightest trace of an anticipation of Quaternions is contained. Endeavouring to extend to *space* the form $a + b\sqrt{-1}$ for the plane, he is guided by analogy to write a directed unit-line in space the form

$$p \cos \alpha + q \cos \beta + r \cos \gamma,$$

where α, β, γ are its inclinations to the three axes. He perceives easily that p, q, r must be *non-reals* : but, he asks, 'seraient-elles imaginaires réductibles à la forme générale $A + B\sqrt{-1}$?' This could not be the answer. In fact they are the **i**, **j**, **k** of the Quaternion Calculus [12].

So the French mathematician François-Joseph Servois (1768–1847), was another person who came very close to discovering quaternions. Furthermore, both Tait and Hamilton were apparently unaware of the paper published by Rodrigues.

And it doesn't stop there. Gauss was extremely cautious, and nervous of publishing anything too revolutionary, just in case he was ridiculed by fellow mathematicians. His diaries reveal that he had anticipated non-Euclidean geometry ahead of Nikolai Ivanovich Lobachevsky. And in a short note from his diary in 1819 [13] he reveals that he had identified a method of finding the product of two quadruples (a, b, c, d) and $(\alpha, \beta, \gamma, \delta)$ as:

$$\begin{aligned}
(A, B, C, D) &= (a, b, c, d)(\alpha, \beta, \gamma, \delta) \\
&= (a\alpha - b\beta - c\gamma - d\delta, \ a\beta + b\alpha - c\delta + d\gamma, \\
&\quad\ a\gamma + b\delta + c\alpha - d\beta, \ a\delta - b\gamma + c\beta + d\alpha).
\end{aligned}$$

At first glance, this result does not look like a quaternion product, but transposing the second and third coordinates of the quadruples, and treating them as quaternions, we have

$$\begin{aligned}
(A, B, C, D) &= (a + ci + bj + dk)(\alpha + \gamma i + \beta j + \delta k) \\
&= a\alpha - c\gamma - b\beta - d\delta + a(\gamma i + \beta j + \delta k) \\
&\quad + \alpha(ci + bj + dk), \ (b\delta - d\beta)i + (d\gamma - c\delta)j + (c\beta - b\gamma)k
\end{aligned}$$

which is identical to Hamilton's quaternion product! Furthermore, Gauss also realised that the product was non-commutative. However, he did not publish his findings, and it was left to Hamilton to invent quaternions for himself, publish his results and take the credit.

In 1881 and 1884, Josiah Willard Gibbs, at Yale University, printed his lecture notes on vector analysis for his students. Gibbs had cut the "umbilical cord" between the real and vector parts of a quaternion and raised the 3D vector as an independent object without any imaginary connotations. Gibbs also took on board the ideas of the German mathematician Hermann Günter Grassmann (1809–1877), who had been developing his own ideas for a vectorial system since 1832. Gibbs also defined the scalar and vector products using the relevant parts of the quaternion product. Finally, in 1901, a student of Gibbs, Edwin Bidwell Wilson, published Gibbs' notes in book form: *Vector Analysis* [14], which contains the notation in use today.

Quaternion algebra is definitely imaginary, yet simply by isolating the vector part and ignoring the imaginary rules, Gibbs was able to reveal a new branch of mathematics that exploded into vector analysis.

Hamilton and his supporters were unable to persuade their peers that quaternions could represent vectorial quantities, and eventually, Gibbs' notation won the day, and quaternions faded from the scene.

In recent years, quaternions have been used in quantum mechanics, flight simulation, and by the computer graphics community, where they are used to rotate vectors about an arbitrary axis. In the intervening years, various people have had the opportunity to investigate the algebra, and propose new ways of harnessing its qualities.

So let's look at three ways of annotating a quaternion q:

$$q = s + xi + yj + zk \qquad (4.1)$$
$$q = s + \mathbf{v} \qquad (4.2)$$
$$q = [s, \ \mathbf{v}] \qquad (4.3)$$
$$\text{where } \{s, x, y, z\} \in \mathbb{R}, \quad \mathbf{v} \in \mathbb{R}^3$$
$$\text{and } i^2 = j^2 = k^2 = -1.$$

The difference is rather subtle: In (4.1) we have Hamilton's original definition with its imaginary terms and associated rules. In (4.2) a "+" sign is used to add a scalar to a vector, which seems strange, yet works. In (4.3) we have an ordered pair comprising a scalar and a vector.

Now you may be thinking: How is it possible to have three different definitions for the same object? Well, I would argue that you can call an object whatever you like, so long as they are algebraically identical. For example, matrix notation is used to represent a set of linear equations, and leads to the same results as every-day algebra. Therefore, both systems of notation are equally valid.

Although I have employed the notation in (4.1) and (4.2) in other publications, in this book I have used ordered pairs. So what we need to show is that Hamilton's original definition of a quaternion (4.1), with its scalar and three imaginary terms, can be replaced by an ordered pair (4.3) comprising a scalar and a "modern" vector.

4.3 Defining a Quaternion

Let's start with two quaternions q_a and q_b à la Hamilton:

$$q_a = s_a + x_a i + y_a j + z_a k$$
$$q_b = s_b + x_b i + y_b j + z_b k$$

and the obligatory rules:

$$i^2 = j^2 = k^2 = ijk = -1$$

$$ij = k, \qquad jk = i, \qquad ki = j$$
$$ji = -k, \quad kj = -i, \quad ik = -j.$$

Our objective is to show that q_a and q_b can also be represented by the ordered pairs

$$\left.\begin{array}{l} q_a = [s_a, \ \mathbf{a}] \\ q_b = [s_b, \ \mathbf{b}] \end{array}\right\} \quad \{s_a, s_b\} \in \mathbb{R}, \quad \{\mathbf{a}, \mathbf{b}\} \in \mathbb{R}^3.$$

I have employed square brackets as part of the definition as parentheses are often used to delimit expressions within a quaternion.

The quaternion product $q_a q_b$ expands to

$$\begin{aligned} q_a q_b = [s_a, \ \mathbf{a}][s_b, \ \mathbf{b}] &= (s_a + x_a i + y_a j + z_a k)(s_b + x_b i + y_b j + z_b k) \\ &= (s_a s_b - x_a x_b - y_a y_b - z_a z_b) \\ &\quad + (s_a x_b + s_b x_a + y_a z_b - y_b z_a)i \\ &\quad + (s_a y_b + s_b y_a + z_a x_b - z_b x_a)j \\ &\quad + (s_a z_b + s_b z_a + x_a y_b - x_b y_a)k. \end{aligned} \tag{4.4}$$

Equation (4.4) takes the form of another quaternion, and confirms that the quaternion product is closed.

At this stage, Hamilton turned the imaginary terms i, j, k into unit Cartesian vectors $\mathbf{i}$, $\mathbf{j}$, $\mathbf{k}$ and transformed (4.4) into a vector form. The problem with this approach is that the vectors retain their imaginary roots. Simon Altmann's suggestion is to replace the imaginaries by the ordered pairs:

$$i = [0, \ \mathbf{i}], \quad j = [0, \ \mathbf{j}], \quad k = [0, \ \mathbf{k}]$$

which are themselves quaternions, and called *quaternion units*.

The idea of defining a quaternion in terms of quaternion units is exactly the same as defining a vector in terms of its unit Cartesian vectors. Furthermore, it permits vectors to exist without any imaginary associations.

Let's substitute these quaternion units in (4.4) together with $[1, \ \mathbf{0}] = 1$:

$$\begin{aligned}{} [s_a, \ \mathbf{a}][s_b, \ \mathbf{b}] &= (s_a s_b - x_a x_b - y_a y_b - z_a z_b)[1, \ \mathbf{0}] \\ &\quad + (s_a x_b + s_b x_a + y_a z_b - y_b z_a)[0, \ \mathbf{i}] \\ &\quad + (s_a y_b + s_b y_a + z_a x_b - z_b x_a)[0, \ \mathbf{j}] \\ &\quad + (s_a z_b + s_b z_a + x_a y_b - x_b y_a)[0, \ \mathbf{k}]. \end{aligned} \tag{4.5}$$

Next, we expand (4.5) using previously defined rules:

$$\begin{aligned}
[s_a, \mathbf{a}][s_b, \mathbf{b}] = {} & [s_a s_b - x_a x_b - y_a y_b - z_a z_b, \ \mathbf{0}] \\
& + [0, \ (s_a x_b + s_b x_a + y_a z_b - y_b z_a)\mathbf{i}] \\
& + [0, \ (s_a y_b + s_b y_a + z_a x_b - z_b x_a)\mathbf{j}] \\
& + [0, \ (s_a z_b + s_b z_a + x_a y_b - x_b y_a)\mathbf{k}].
\end{aligned} \tag{4.6}$$

A vertical scan of (4.6) reveals some hidden vectors:

$$\begin{aligned}
[s_a, \mathbf{a}][s_b, \mathbf{b}] = {} & [s_a s_b - x_a x_b - y_a y_b - z_a z_b, \ \mathbf{0}] \\
& + [0, \ s_a(x_b\mathbf{i} + y_b\mathbf{j} + z_b\mathbf{k}) + s_b(x_a\mathbf{i} + y_a\mathbf{j} + z_a\mathbf{k}) \\
& + (y_a z_b - y_b z_a)\mathbf{i} + (z_a x_b - z_b x_a)\mathbf{j} + (x_a y_b - x_b y_a)\mathbf{k}].
\end{aligned} \tag{4.7}$$

Equation (4.7) contains two ordered pairs which can now be combined:

$$\begin{aligned}
[s_a, \mathbf{a}][s_b, \mathbf{b}] = {} & [s_a s_b - x_a x_b - y_a y_b - z_a z_b, \\
& s_a(x_b\mathbf{i} + y_b\mathbf{j} + z_b\mathbf{k}) + s_b(x_a\mathbf{i} + y_a\mathbf{j} + z_a\mathbf{k}) \\
& + (y_a z_b - y_b z_a)\mathbf{i} + (z_a x_b - z_b x_a)\mathbf{j} + (x_a y_b - x_b y_a)\mathbf{k}].
\end{aligned} \tag{4.8}$$

If we make

$$\mathbf{a} = x_a\mathbf{i} + y_a\mathbf{j} + z_a\mathbf{k}$$
$$\mathbf{b} = x_b\mathbf{i} + y_b\mathbf{j} + z_b\mathbf{k}$$

and substitute them in (4.8) we get:

$$[s_a, \mathbf{a}][s_b, \mathbf{b}] = [s_a s_b - \mathbf{a} \cdot \mathbf{b}, \ s_a\mathbf{b} + s_b\mathbf{a} + \mathbf{a} \times \mathbf{b}] \tag{4.9}$$

which defines the quaternion product.

From now on, we don't have to worry about Hamilton's rules as they are embedded within (4.9). Furthermore, our vectors have no imaginary associations.

Although Rodrigues did not have access to Gibbs' vector notation used in (4.9), he managed to calculate the equivalent algebraic expression, which was some achievement.

4.3.1 The Quaternion Units

Using (4.9) we can check to see if the quaternion units are imaginary by squaring them:

$$i = [0, \ \mathbf{i}]$$
$$i^2 = [0, \ \mathbf{i}][0, \ \mathbf{i}]$$
$$= [\mathbf{i} \cdot \mathbf{i}, \ \mathbf{i} \times \mathbf{i}]$$
$$= [-1, \ \mathbf{0}]$$

which is a *real quaternion* and equivalent to -1, confirming that $[0, \ \mathbf{i}]$ is imaginary. Using a similar expansion we can shown that $[0, \ \mathbf{j}]$ and $[0, \ \mathbf{k}]$ have the same property.

Now let's compute the products ij, jk and ki:

$$ij = [0, \ \mathbf{i}][0, \ \mathbf{j}]$$
$$= [-\mathbf{i} \cdot \mathbf{j}, \ \mathbf{i} \times \mathbf{j}]$$
$$= [0, \ \mathbf{k}]$$

which is the quaternion unit k.

$$jk = [0, \ \mathbf{j}][0, \ \mathbf{k}]$$
$$= [-\mathbf{j} \cdot \mathbf{k}, \ \mathbf{j} \times \mathbf{k}]$$
$$= [0, \ \mathbf{i}]$$

which is the quaternion unit i.

$$ki = [0, \ \mathbf{k}][0, \ \mathbf{i}]$$
$$= [-\mathbf{k} \cdot \mathbf{i}, \ \mathbf{k} \times \mathbf{i}]$$
$$= [0, \ \mathbf{j}]$$

which is the quaternion unit j.

Next, let's confirm that $ijk = -1$:

$$ijk = [0, \ \mathbf{i}][0, \ \mathbf{j}][0, \ \mathbf{k}]$$
$$= [0, \ \mathbf{k}][0, \ \mathbf{k}]$$
$$= [-\mathbf{k} \cdot \mathbf{k}, \ \mathbf{k} \times \mathbf{k}]$$
$$= [-1, \ \mathbf{0}]$$

which is a real quaternion equivalent to -1, confirming that $ijk = -1$.

Thus the notation of ordered pairs upholds all of Hamilton's rules. However, the last double product assumes that quaternions are associative. So let's double check to show that $(ij)k = i(jk)$:

$$i(jk) = [0, \mathbf{i}][0, \mathbf{j}][0, \mathbf{k}]$$
$$= [0, \mathbf{i}][0, \mathbf{i}]$$
$$= [-\mathbf{i} \cdot \mathbf{i}, \ \mathbf{i} \times \mathbf{i}]$$
$$= [-1, \ \mathbf{0}]$$

which is correct.

Although we have yet to discover how quaternions are used to rotate vectors, let's concentrate on their algebraic traits by evaluating an example.

$$q_a = [1, \ 2\mathbf{i} + 3\mathbf{j} + 4\mathbf{k}]$$
$$q_b = [2, \ 3\mathbf{i} + 4\mathbf{j} + 5\mathbf{k}]$$

their product is

$$
\begin{aligned}
q_a q_b &= [1, \ 2\mathbf{i} + 3\mathbf{j} + 4\mathbf{k}][2, \ 3\mathbf{i} + 4\mathbf{j} + 5\mathbf{k}] \\
&= [1 \times 2 - (2 \times 3 + 3 \times 4 + 4 \times 5), \\
&\quad 1(3\mathbf{i} + 4\mathbf{j} + 5\mathbf{k}) + 2(2\mathbf{i} + 3\mathbf{j} + 4\mathbf{k}) \\
&\quad + (3 \times 5 - 4 \times 4)\mathbf{i} - (2 \times 5 - 4 \times 3)\mathbf{j} + (2 \times 4 - 3 \times 3)\mathbf{k}] \\
&= [-36, \ 7\mathbf{i} + 10\mathbf{j} + 13\mathbf{k} - \mathbf{i} + 2\mathbf{j} - \mathbf{k}] \\
&= [-36, \ 6\mathbf{i} + 12\mathbf{j} + 12\mathbf{k}]
\end{aligned}
$$

which is another ordered pair representing a quaternion.

Having shown that Hamilton's *imaginary* notation has a vector equivalent, and can be represented as an ordered pair, we continue with this notation and describe other features of quaternions. Note that we can abandon Hamilton's rules as they are embedded within the definition of the quaternion product, and will surface in the following definitions.

4.4 Algebraic Definition

A quaternion is the ordered pair:

$$q = [s, \ \mathbf{v}], \quad s \in \mathbb{R}, \quad \mathbf{v} \in \mathbb{R}^3.$$

If we express $\mathbf{v}$ in terms of its components, we have

$$q = [s, \ x\mathbf{i} + y\mathbf{j} + z\mathbf{k}], \quad \{s, x, y, z\} \in \mathbb{R}, \quad \{\mathbf{i}, \mathbf{j}, \mathbf{k}\} \in \mathbb{R}^3.$$

4.5   Adding and Subtracting Quaternions

Addition and subtraction employ the following rule:

$$q_a = [s_a, \ \mathbf{a}]$$
$$q_b = [s_b, \ \mathbf{b}]$$
$$q_a \pm q_b = [s_a \pm s_b, \ \mathbf{a} \pm \mathbf{b}].$$

For example,

$$q_a = [0.5, \ 2\mathbf{i} + 3\mathbf{j} - 4\mathbf{k}]$$
$$q_b = [0.1, \ 4\mathbf{i} + 5\mathbf{j} + 6\mathbf{k}]$$
$$q_a + q_b = [0.6, \ 6\mathbf{i} + 8\mathbf{j} + 2\mathbf{k}]$$
$$q_a - q_b = [0.4, \ -2\mathbf{i} - 2\mathbf{j} - 10\mathbf{k}].$$

4.6   Real Quaternion

A *real quaternion* has a zero vector term:

$$q = [s, \ \mathbf{0}].$$

The product of two real quaternions is

$$q_a = [s_a, \ \mathbf{0}]$$
$$q_b = [s_b, \ \mathbf{0}]$$
$$q_a q_b = [s_a, \ \mathbf{0}][s_b, \ \mathbf{0}]$$
$$= [s_a s_b, \ \mathbf{0}]$$

which is another real quaternion, and shows that they behave just like real numbers:

$$[s, \ \mathbf{0}] \equiv s.$$

We have already come across this with complex numbers containing a zero imaginary term:

$$a + bi = a, \quad \text{when } b = 0.$$

4.7 Scaling a Quaternion

Intuition suggests that multiplying a quaternion by a scalar should obey the rule:

$$q = [s, \ \mathbf{v}]$$
$$\lambda q = \lambda[s, \ \mathbf{v}], \quad \lambda \in \mathbb{R}$$
$$= [\lambda s, \ \lambda \mathbf{v}].$$

We can confirm our intuition by multiplying a quaternion by a scalar in the form of a real quaternion:

$$q = [s, \ \mathbf{v}]$$
$$\lambda = [\lambda, \ \mathbf{0}]$$
$$\lambda q = [\lambda, \ \mathbf{0}][s, \ \mathbf{v}]$$
$$= [\lambda s, \ \lambda \mathbf{v}]$$

which is excellent confirmation.

4.8 Pure Quaternion

Hamilton defined a *pure quaternion* as one having a zero scalar term:

$$q = xi + yj + zk$$

and was just a vector, with all its imaginary qualities. However, Simon Altmann and others, believe that this was a serious mistake on Hamilton's part to call a quaternion with a zero real term, a vector.

The main issue is that there are two types of vectors: *polar* and *axial*, also called a *pseudovector*. Richard Feynman describes polar vectors as "honest" vectors [15] and represent the every-day vectors of directed lines. Whereas, axial vectors are computed from polar vectors, such as in a vector product. However, these two types of vector do not behave in the same way when transformed. For example, given two "honest", polar vectors $\mathbf{a}$ and $\mathbf{b}$, we can compute the axial vector: $\mathbf{c} = \mathbf{a} \times \mathbf{b}$. Next, if we subject $\mathbf{a}$ and $\mathbf{b}$ to an inversion transform through the origin, such that $\mathbf{a}$ becomes $-\mathbf{a}$, and $\mathbf{b}$ becomes $-\mathbf{b}$, and compute their cross product $(-\mathbf{a}) \times (-\mathbf{b})$, we still get $\mathbf{c}$! Which implies that the axial vector $\mathbf{c}$ must not be transformed along with $\mathbf{a}$ and $\mathbf{b}$. It could be argued that the inversion transform is not a "proper" transform as it turns a right-handed set of axes into a left-handed set. Unfortunately, Hamilton was not aware of this distinction, as he had only just invented vectors. However, in the intervening years, it has become evident that Hamilton's quaternion vector is an axial vector, and not a polar vector.

As we will see, in 3D rotations quaternions take the form

$$q = \left[\cos\left(\tfrac{\theta}{2}\right),\ \sin\left(\tfrac{\theta}{2}\right)\mathbf{v} \right]$$

where θ is the angle of rotation and $\mathbf{v}$ is the axis of rotation, and when we set $\theta = 180°$, we get

$$q = [0,\ \mathbf{v}]$$

which remains a quaternion, even though it only contains a vector part.

Consequently, we define a *pure quaternion* as

$$q = [0,\ \mathbf{v}].$$

The product of two pure quaternions is

$$q_a = [0,\ \mathbf{a}]$$
$$q_b = [0,\ \mathbf{b}]$$
$$q_a q_b = [0,\ \mathbf{a}][0,\ \mathbf{b}]$$
$$= [-\mathbf{a} \cdot \mathbf{b},\ \mathbf{a} \times \mathbf{b}]$$

which is no longer "pure", as some of the original vector information has "tunnelled" across into the real part via the dot product.

4.9   Unit Quaternion

Let's pursue this analysis further by introducing some familiar vector notation.

Given vector $\mathbf{v}$, then

$$\mathbf{v} = v\hat{\mathbf{v}}, \quad \text{where} \quad v = |\mathbf{v}|, \quad |\hat{\mathbf{v}}| = 1.$$

Combining this with the definition of a pure quaternion we get:

$$q = [0,\ \mathbf{v}]$$
$$= [0,\ v\hat{\mathbf{v}}]$$
$$= v[0,\ \hat{\mathbf{v}}]$$

and reveals the object $[0,\ \hat{\mathbf{v}}]$ which is called the *unit quaternion* and comprises a zero scalar and a unit vector. It is usual to identify this unit quaternion as $\hat{q}$:

$$\hat{q} = [0,\ \hat{\mathbf{v}}].$$

So now we have a notation similar to that of vectors where a vector **v** is described in terms of its unit form:

$$\mathbf{v} = v\hat{\mathbf{v}}$$

and a quaternion q is also described in terms of its unit form:

$$q = v\hat{q}.$$

Note that $\hat{q}$ is imaginary:

$$\begin{aligned}
\hat{q}^2 &= [0, \ \hat{\mathbf{v}}][0, \ \hat{\mathbf{v}}] \\
&= [-\hat{\mathbf{v}} \cdot \hat{\mathbf{v}}, \ \hat{\mathbf{v}} \times \hat{\mathbf{v}}] \\
&= [-1, \ \mathbf{0}] \\
&= -1
\end{aligned}$$

which is not too surprising, bearing in mind Hamilton's original invention!

4.10 Additive Form of a Quaternion

We now come to the idea of splitting a quaternion into its constituent parts: a real quaternion and a pure quaternion. Again, intuition suggests that we can write a quaternion as

$$\begin{aligned}
q &= [s, \ \mathbf{v}] \\
&= [s, \ \mathbf{0}] + [0, \ \mathbf{v}]
\end{aligned}$$

and we can test this by forming the algebraic product of two quaternions represented in this way:

$$\begin{aligned}
q_a &= [s_a, \ \mathbf{0}] + [0, \ \mathbf{a}] \\
q_b &= [s_b, \ \mathbf{0}] + [0, \ \mathbf{b}] \\
q_a q_b &= \big([s_a, \ \mathbf{0}] + [0, \ \mathbf{a}]\big)\big([s_b, \ \mathbf{0}] + [0, \ \mathbf{b}]\big) \\
&= [s_a, \ \mathbf{0}][s_b, \ \mathbf{0}] + [s_a, \ \mathbf{0}][0, \ \mathbf{b}] + [0, \ \mathbf{a}][s_b, \ \mathbf{0}] + [0, \ \mathbf{a}][0, \ \mathbf{b}] \\
&= [s_a s_b, \ \mathbf{0}] + [0, \ s_a\mathbf{b}] + [0, \ s_b\mathbf{a}] + [-\mathbf{a} \cdot \mathbf{b}, \ \mathbf{a} \times \mathbf{b}] \\
&= [s_a s_b - \mathbf{a} \cdot \mathbf{b}, \ s_a\mathbf{b} + s_b\mathbf{a} + \mathbf{a} \times \mathbf{b}]
\end{aligned}$$

which is correct, and confirms that the additive form works.

4.11   Binary Form of a Quaternion

Having shown that the additive form of a quaternion works, and discovered the unit quaternion, we can join the two objects together as follows:

$$q = [s, \ \mathbf{v}]$$
$$= [s, \ \mathbf{0}] + [0, \ \mathbf{v}]$$
$$= [s, \ \mathbf{0}] + v[0, \ \hat{\mathbf{v}}]$$
$$= s + v\hat{q}.$$

Just to recap, s is a scalar, v is the length of the vector term, and $\hat{q}$ is the unit quaternion $[0, \ \hat{\mathbf{v}}]$.

Look how similar this notation is to a complex number:

$$\left. \begin{array}{l} z = a + bi \\ q = s + v\hat{q} \end{array} \right\} \quad \{a, b, s, v\} \in \mathbb{R}, \quad i \in \mathbb{I}$$

and $\hat{q}$ is the unit quaternion.

4.12   Quaternion Conjugate

We have already discovered that the conjugate of a complex number $z = a + bi$ is given by

$$\bar{z} = a - bi$$

and is very useful in computing the inverse of z. The *quaternion conjugate* plays a similar role in computing the inverse of a quaternion. Therefore, given

$$q = [s, \ \mathbf{v}]$$

the quaternion conjugate is defined as

$$\bar{q} = [s, \ -\mathbf{v}]$$

If we compute the product $q\overline{q}$ we obtain

$$
\begin{aligned}
q\overline{q} &= [s, \ \mathbf{v}][s, \ -\mathbf{v}] \\
&= \left[s^2 - \mathbf{v} \cdot (-\mathbf{v}), \ -s\mathbf{v} + s\mathbf{v} + \mathbf{v} \times (-\mathbf{v})\right] \\
&= \left[s^2 + \mathbf{v} \cdot \mathbf{v}, \ \mathbf{0}\right] \\
&= \left[s^2 + v^2, \ \mathbf{0}\right].
\end{aligned}
$$

Let's show that $q\overline{q} = \overline{q}q$:

$$
\begin{aligned}
\overline{q}q &= [s, \ -\mathbf{v}][s, \ \mathbf{v}] \\
&= \left[s^2 - (-\mathbf{v}) \cdot \mathbf{v}, \ s\mathbf{v} - s\mathbf{v} + (-\mathbf{v}) \times \mathbf{v}\right] \\
&= \left[s^2 + \mathbf{v} \cdot \mathbf{v}, \ \mathbf{0}\right] \\
&= \left[s^2 + v^2, \ \mathbf{0}\right] \\
&= q\overline{q}.
\end{aligned}
$$

Now let's show that $\overline{q_a q_b} = \overline{q}_b \overline{q}_a$:

$$
\begin{aligned}
q_a &= [s_a, \ \mathbf{a}] \\
q_b &= [s_b, \ \mathbf{b}] \\
q_a q_b &= [s_a, \ \mathbf{a}][s_b, \ \mathbf{b}] \\
&= [s_a s_b - \mathbf{a} \cdot \mathbf{b}, \ s_a \mathbf{b} + s_b \mathbf{a} + \mathbf{a} \times \mathbf{b}] \\
\overline{q_a q_b} &= [s_a s_b - \mathbf{a} \cdot \mathbf{b}, \ -s_a \mathbf{b} - s_b \mathbf{a} - \mathbf{a} \times \mathbf{b}].
\end{aligned}
\qquad (4.10)
$$

Next, we compute $\overline{q}_b \overline{q}_a$:

$$
\begin{aligned}
\overline{q}_a &= [s_a, \ -\mathbf{a}] \\
\overline{q}_b &= [s_b, \ -\mathbf{b}] \\
\overline{q}_b \overline{q}_a &= [s_b, \ -\mathbf{b}][s_a, \ -\mathbf{a}] \\
&= [s_a s_b - \mathbf{a} \cdot \mathbf{b}, \ -s_a \mathbf{b} - s_b \mathbf{a} - \mathbf{a} \times \mathbf{b}].
\end{aligned}
\qquad (4.11)
$$

And as (4.10) equals (4.11), $\overline{q_a q_b} = \overline{q}_b \overline{q}_a$.

4.13   Norm of a Quaternion

The *norm* of a complex number $z = a + bi$ is defined as

$$
\|z\| = \sqrt{a^2 + b^2}
$$

which allows us to write

$$z\bar{z} = \|z\|^2.$$

Similarly, the norm of a quaternion $q = [s, \ \mathbf{v}]$ is defined as

$$\|q\| = \sqrt{s^2 + v^2}$$

which allows us to write

$$q\bar{q} = \|q\|^2.$$

For example,

$$q = [1, \ 4\mathbf{i} + 4\mathbf{j} - 4\mathbf{k}]$$
$$\|q\| = \sqrt{1^2 + 4^2 + 4^2 + (-4)^2}$$
$$= \sqrt{49}$$
$$= 7.$$

4.14   Normalised Quaternion

A quaternion with a unit norm is called a *normalised quaternion*. For example, the quaternion $q = [s, \ \mathbf{v}]$ is *normalised* by dividing it by $\|q\|$:

$$q' = \frac{q}{\sqrt{s^2 + v^2}}.$$

We must be careful not to confuse the unit quaternion with a unit-norm quaternion. The unit quaternion is $[0, \ \hat{\mathbf{v}}]$ with a unit-vector part, whereas a unit-norm quaternion is normalised such that $s^2 + v^2 = 1$.

I will be careful to distinguish between these two terms as many authors – including myself – use the term unit quaternion to describe a quaternion with a unit norm. For example

$$q = [1, \ 4\mathbf{i} + 4\mathbf{j} - 4\mathbf{k}]$$

has a norm of 7, and q is normalised by dividing by 7:

$$q' = \tfrac{1}{7} [1, \ 4\mathbf{i} + 4\mathbf{j} - 4\mathbf{k}].$$

The type of unit-norm quaternion we will be using takes the form:

$$q = \left[\cos\left(\tfrac{\theta}{2}\right),\ \sin\left(\tfrac{\theta}{2}\right) \hat{\mathbf{v}} \right]$$

because $\cos^2\left(\tfrac{\theta}{2}\right) + \sin^2\left(\tfrac{\theta}{2}\right) = 1$.

4.15 Quaternion Products

Having shown that ordered pairs can represent a quaternion and its various manifestations, let's summarise the products we will eventually encounter. To start, we have the product of two normal quaternions:

$$
\begin{aligned}
q_a q_b &= [s_a,\ \mathbf{a}][s_b,\ \mathbf{b}] \\
&= [s_a s_b - \mathbf{a} \cdot \mathbf{b},\ s_a \mathbf{b} + s_b \mathbf{a} + \mathbf{a} \times \mathbf{b}].
\end{aligned}
$$

4.15.1 Pure Quaternion Product

Given two pure quaternions:

$$
\begin{aligned}
q_a &= [0,\ \mathbf{a}] \\
q_b &= [0,\ \mathbf{b}] \\
q_a q_b &= [0,\ \mathbf{a}][0,\ \mathbf{b}] \\
&= [-\mathbf{a} \cdot \mathbf{b},\ \mathbf{a} \times \mathbf{b}].
\end{aligned}
$$

4.15.2 Unit-Norm Quaternion Product

Given two unit-norm quaternions:

$$
\begin{aligned}
q_a &= [s_a,\ \mathbf{a}] \\
q_b &= [s_b,\ \mathbf{b}] \\
a &= |\mathbf{a}| \\
b &= |\mathbf{b}|
\end{aligned}
$$

where $|q_a| = |q_b| = 1$, their product is another unit-norm quaternion, which is proved as follows.

Fig. 4.2 The geometry for c

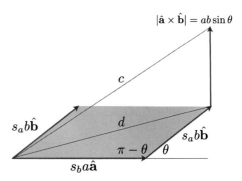

We assume $q_c = [s_c, \ \mathbf{c}]$ and show that $|q_c| = s_c^2 + c^2 = 1$, where $c = |\mathbf{c}|$ and

$$[s_c, \ \mathbf{c}] = [s_a, \ \mathbf{a}][s_b, \ \mathbf{b}]$$
$$= [s_a s_b - \mathbf{a} \cdot \mathbf{b}, \ s_a \mathbf{b} + s_b \mathbf{a} + \mathbf{a} \times \mathbf{b}].$$

Let's assume the angle between $\mathbf{a}$ and $\mathbf{b}$ is θ, which permits us to write

$$s_c = s_a s_b - ab \cos \theta$$
$$\mathbf{c} = s_a b \hat{\mathbf{b}} + s_b a \hat{\mathbf{a}} + ab \sin \theta \left(\hat{\mathbf{a}} \times \hat{\mathbf{b}} \right).$$

Therefore,

$$s_c^2 = (s_a s_b - ab \cos \theta)(s_a s_b - ab \cos \theta)$$
$$= s_a^2 s_b^2 - 2 s_a s_b ab \cos \theta + a^2 b^2 \cos^2 \theta.$$

Figure 4.2 shows the geometry representing $\mathbf{c}$.

$$d^2 = s_b^2 a^2 + s_a^2 b^2 - 2 s_a s_b ab \cos(\pi - \theta)$$
$$= s_b^2 a^2 + s_a^2 b^2 + 2 s_a s_b ab \cos \theta$$
$$c^2 = d^2 + a^2 b^2 \sin^2 \theta$$
$$= s_b^2 a^2 + s_a^2 b^2 + 2 s_a s_b ab \cos \theta + a^2 b^2 \sin^2 \theta$$
$$s_c^2 + c^2 = s_a^2 s_b^2 - 2 s_a s_b ab \cos \theta + a^2 b^2 \cos^2 \theta + s_b^2 a^2 + s_a^2 b^2 + 2 s_a s_b ab \cos \theta + a^2 b^2 \sin^2 \theta$$
$$= s_a^2 s_b^2 + a^2 b^2 + s_b^2 a^2 + s_a^2 b^2$$
$$= s_a^2 \left(s_b^2 + b^2 \right) + a^2 \left(s_b^2 + b^2 \right)$$
$$= s_a^2 + a^2$$
$$= 1.$$

Therefore, the product of two unit-norm quaternions is another unit-norm quaternion. Consequently, multiplying a quaternion by a unit-norm quaternion, does not change its norm:

$$q_a = [s_a, \ \mathbf{a}]$$
$$\|q_a\| = 1$$
$$q_b = [s_b, \ \mathbf{b}]$$
$$\|q_a q_b\| = \|q_b\|.$$

4.15.3   Square of a Quaternion

The square of a quaternion is given by

$$q = [s, \ \mathbf{v}]$$
$$q^2 = [s, \ \mathbf{v}][s, \ \mathbf{v}]$$
$$= \left[s^2 - \mathbf{v} \cdot \mathbf{v}, \ 2s\mathbf{v} + \mathbf{v} \times \mathbf{v}\right]$$
$$= \left[s^2 - \mathbf{v} \cdot \mathbf{v}, \ 2s\mathbf{v}\right]$$
$$= \left[s^2 - x^2 - y^2 - z^2, \ 2s(x\mathbf{i} + y\mathbf{j} + z\mathbf{k})\right].$$

For example:

$$q = [7, \ 2\mathbf{i} + 3\mathbf{j} + 4\mathbf{k}]$$
$$q^2 = \left[7^2 - 2^2 - 3^2 - 4^2, \ 14(2\mathbf{i} + 3\mathbf{j} + 4\mathbf{k})\right]$$
$$= [20, \ 28\mathbf{i} + 42\mathbf{j} + 56\mathbf{k}].$$

The square of a pure quaternion is

$$q = [0, \ \mathbf{v}]$$
$$q^2 = [0, \ \mathbf{v}][0, \ \mathbf{v}]$$
$$= [0 - \mathbf{v} \cdot \mathbf{v}, \ \mathbf{v} \times \mathbf{v}]$$
$$= [0 - \mathbf{v} \cdot \mathbf{v}, \ \mathbf{0}]$$
$$= \left[-\left(x^2 + y^2 + z^2\right), \ \mathbf{0}\right]$$

which makes the square of a pure, unit-norm quaternion equal to -1, and was one of the results, to which some 19th-century mathematicians objected.

4.15.4 Norm of the Quaternion Product

In proving that the product of two unit-norm quaternions is another unit-norm quaternion we saw that

$$q_a = [s_a, \ \mathbf{a}]$$
$$q_b = [s_b, \ \mathbf{b}]$$
$$q_c = q_a q_b$$
$$\|q_c\|^2 = s_a^2 \left(s_b^2 + b^2\right) + a^2 \left(s_b^2 + b^2\right)$$
$$= \left(s_a^2 + a^2\right)\left(s_b^2 + b^2\right)$$

which, if we ignore the constraint of unit-norm quaternions, shows that the norm of a quaternion product equals the product of the individual norms:

$$\|q_a q_b\|^2 = \|q_a\|^2 \|q_b\|^2$$
$$\|q_a q_b\| = \|q_a\| \ \|q_b\|.$$

4.16 Inverse Quaternion

An important feature of quaternion algebra is the ability to divide two quaternions q_b/q_a, as long as q_a does not vanish.

By definition, the inverse q^{-1} of q satisfies

$$qq^{-1} = [1, \ \mathbf{0}] = 1. \tag{4.12}$$

To isolate q^{-1}, we multiply (4.12) by $\overline{q}$

$$\overline{q}qq^{-1} = \overline{q}$$
$$\|q\|^2 q^{-1} = \overline{q} \tag{4.13}$$

and from (4.13) we can write

$$q^{-1} = \frac{\overline{q}}{\|q\|^2}.$$

If q is a unit-norm quaternion, then

$$q^{-1} = \overline{q}$$

which is useful in the context of rotations. Furthermore, as

$$\overline{q_a q_b} = \overline{q}_b \overline{q}_a$$

then

$$(q_a q_b)^{-1} = q_b^{-1} q_a^{-1}.$$

Note that $qq^{-1} = q^{-1}q$:

$$qq^{-1} = \frac{q\overline{q}}{\|q\|^2} = 1$$

$$q^{-1}q = \frac{\overline{q}q}{\|q\|^2} = 1.$$

Thus, we represent the quotient q_b/q_a as

$$q_c = \frac{q_b}{q_a}$$
$$= q_b q_a^{-1}$$
$$= \frac{q_b \overline{q}_a}{\|q_a\|^2}.$$

For completeness let's evaluate the inverse of q where

$$q = \left[1, \ \tfrac{1}{\sqrt{3}}\mathbf{i} + \tfrac{1}{\sqrt{3}}\mathbf{j} + \tfrac{1}{\sqrt{3}}\mathbf{k} \right]$$
$$\overline{q} = \left[1, \ -\tfrac{1}{\sqrt{3}}\mathbf{i} - \tfrac{1}{\sqrt{3}}\mathbf{j} - \tfrac{1}{\sqrt{3}}\mathbf{k} \right]$$
$$\|q\|^2 = 1 + \tfrac{1}{3} + \tfrac{1}{3} + \tfrac{1}{3} = 2$$
$$q^{-1} = \frac{\overline{q}}{\|q\|^2} = \tfrac{1}{2}\left[1, \ -\tfrac{1}{\sqrt{3}}\mathbf{i} - \tfrac{1}{\sqrt{3}}\mathbf{j} - \tfrac{1}{\sqrt{3}}\mathbf{k} \right].$$

It should be clear that $q^{-1}q = 1$:

$$q^{-1}q = \tfrac{1}{2}\left[1, \ -\tfrac{1}{\sqrt{3}}\mathbf{i} - \tfrac{1}{\sqrt{3}}\mathbf{j} - \tfrac{1}{\sqrt{3}}\mathbf{k} \right]\left[1, \ \tfrac{1}{\sqrt{3}}\mathbf{i} + \tfrac{1}{\sqrt{3}}\mathbf{j} + \tfrac{1}{\sqrt{3}}\mathbf{k} \right]$$
$$= \tfrac{1}{2}\left[1 + \tfrac{1}{3} + \tfrac{1}{3} + \tfrac{1}{3}, \ \mathbf{0} \right]$$
$$= 1.$$

4.17   Quaternion Matrix

Matrices provide another way to express a quaternion product. For convenience, let's repeat (4.8) again and show it in matrix form:

$$[s_a, \ \mathbf{a}][s_b, \ \mathbf{b}] = [s_a s_b - x_a x_b - y_a y_b - z_a z_b,$$
$$s_a(x_b\mathbf{i} + y_b\mathbf{j} + z_b\mathbf{k}) + s_b(x_a\mathbf{i} + y_a\mathbf{j} + z_a\mathbf{k}) +$$
$$(y_a z_b - y_b z_a)\mathbf{i} + (z_a x_b - z_b x_a)\mathbf{j} + (x_a y_b - x_b y_a)\mathbf{k}]$$

$$= \begin{bmatrix} s_a & -x_a & -y_a & -z_a \\ x_a & s_a & -z_a & y_a \\ y_a & z_a & s_a & -x_a \\ z_a & -y_a & x_a & s_a \end{bmatrix} \begin{bmatrix} s_b \\ x_b \\ y_b \\ z_b \end{bmatrix}. \tag{4.14}$$

Let's recompute the product $q_a q_b$ using the above matrix:

$$q_a = [1, \ 2\mathbf{i} + 3\mathbf{j} + 4\mathbf{k}]$$
$$q_b = [2, \ 3\mathbf{i} + 4\mathbf{j} + 5\mathbf{k}]$$

$$q_a q_b = \begin{bmatrix} 1 & -2 & -3 & -4 \\ 2 & 1 & -4 & 3 \\ 3 & 4 & 1 & -2 \\ 4 & -3 & 2 & 1 \end{bmatrix} \begin{bmatrix} 2 \\ 3 \\ 4 \\ 5 \end{bmatrix}$$

$$= \begin{bmatrix} -36 \\ 6 \\ 12 \\ 12 \end{bmatrix}$$

$$= [-36, \ 6\mathbf{i} + 12\mathbf{j} + 12\mathbf{k}].$$

4.17.1   Orthogonal Quaternion Matrix

We can demonstrate that the unit-norm quaternion matrix is orthogonal by showing that the product with its transpose equals the identity matrix. As we are dealing with matrices, Q will represent the matrix for q:

$$q = [s, \ x\mathbf{i} + y\mathbf{j} + z\mathbf{k}]$$

where $\quad 1 = s^2 + x^2 + y^2 + z^2$

$$Q = \begin{bmatrix} s & -x & -y & -z \\ x & s & -z & y \\ y & z & s & -x \\ z & -y & x & s \end{bmatrix}$$

$$Q^T = \begin{bmatrix} s & x & y & z \\ -x & s & z & -y \\ -y & -z & s & x \\ -z & y & -x & s \end{bmatrix}$$

$$QQ^T = \begin{bmatrix} s & -x & -y & -z \\ x & s & -z & y \\ y & z & s & -x \\ z & -y & x & s \end{bmatrix} \begin{bmatrix} s & x & y & z \\ -x & s & z & -y \\ -y & -z & s & x \\ -z & y & -x & s \end{bmatrix}$$

$$= \begin{bmatrix} 1 & 0 & 0 & 0 \\ 0 & 1 & 0 & 0 \\ 0 & 0 & 1 & 0 \\ 0 & 0 & 0 & 1 \end{bmatrix}$$

For this to occur, $Q^T = Q^{-1}$.

4.18 Quaternion Algebra

Ordered pairs provide a simple notation for representing quaternions, and allow us to represent the real unit 1 as $[1, \ \mathbf{0}]$, and the imaginary units i, j, k as $[0, \ \mathbf{i}]$, $[0, \ \mathbf{j}]$, $[0, \ \mathbf{k}]$ respectively. A quaternion then becomes a linear combination of these elements with associated real coefficients. Under such conditions, the elements form the *basis* for an algebra over the field of reals.

Furthermore, because quaternion algebra supports division, and obeys the normal axioms of algebra, except that multiplication is non-commutative, it is called a *division algebra*. Ferdinand Georg Frobenius proved in 1878 that only three such real associative division algebras exist: real numbers, complex numbers and quaternions [3].

The Cayley numbers $\mathbb{O}$, constitute a real division algebra, but the Cayley numbers are 8-dimensional and are not associative, i.e. $a(bc) \neq (ab)c$ for all $\{a, b, c\} \in \mathbb{O}$.

4.19   Summary

Quaternions are very similar to complex numbers, apart from the fact that they
have three imaginary units, rather than one. Consequently, they inherit some of the
properties associated with complex numbers, such as norm, conjugate, unit norm
and inverse. They can also be added, subtracted, multiplied and divided. However,
unlike complex numbers, they anti-commute when multiplied.

4.19.1   Summary of Operations

Quaternion

$$q_a = [s_a, \ \mathbf{a}] = [s_a, \ x_a\mathbf{i} + y_a\mathbf{j} + z_a\mathbf{k}]$$
$$q_b = [s_b, \ \mathbf{b}] = [s_b, \ x_b\mathbf{i} + y_b\mathbf{j} + z_b\mathbf{k}].$$

Adding and subtracting

$$q_a \pm q_b = [s_a \pm s_b, \ \mathbf{a} \pm \mathbf{b}].$$

Product

$$q_a q_b = [s_a, \ \mathbf{a}][s_b, \ \mathbf{b}]$$
$$= [s_a s_b - \mathbf{a} \cdot \mathbf{b}, \ s_a\mathbf{b} + s_b\mathbf{a} + \mathbf{a} \times \mathbf{b}]$$
$$= \begin{bmatrix} s_a & -x_a & -y_a & -z_a \\ x_a & s_a & -z_a & y_a \\ y_a & z_a & s_a & -x_a \\ z_a & -y_a & x_a & s_a \end{bmatrix} \begin{bmatrix} s_b \\ x_b \\ y_b \\ z_b \end{bmatrix}.$$

Square

$$q^2 = [s, \ \mathbf{v}][s, \ \mathbf{v}]$$
$$= \left[s^2 - x^2 - y^2 - z^2, \ 2s(x\mathbf{i} + y\mathbf{j} + z\mathbf{k}) \right].$$

Pure

$$q^2 = [0, \ \mathbf{v}][0, \ \mathbf{v}]$$
$$= \left[-(x^2 + y^2 + z^2), \ \mathbf{0} \right].$$

Norm

$$\|q\| = \sqrt{s^2 + v^2}.$$

Norm of the Product

$$\|q_a q_b\|^2 = \|q_a\|^2 \|q_b\|^2$$
$$\|q_a q_b\| = \|q_a\| \, \|q_b\|.$$

Unit norm

$$\|q\| = \sqrt{s^2 + v^2} = 1.$$

Conjugate

$$\overline{q} = [s, \ -\mathbf{v}]$$
$$(\overline{q_a q_b}) = \overline{q}_b \overline{q}_a.$$

Inverse

$$q^{-1} = \frac{\overline{q}}{\|q\|^2}$$
$$(q_a q_b)^{-1} = q_b^{-1} q_a^{-1}.$$

4.20 Worked Examples

Here are some further worked examples that employ the ideas described above. In some cases, a test is included to confirm the result.

4.20.1 Adding and Subtracting Quaternions

Add and subtract q_a and q_b.

$$q_a = [2, \ -2\mathbf{i} + 3\mathbf{j} - 4\mathbf{k}], \qquad q_b = [1, \ -2\mathbf{i} + 5\mathbf{j} - 6\mathbf{k}].$$

Solution: Add and subtract the real and vector elements.

$$q_a + q_b = [3, \ -4\mathbf{i} + 8\mathbf{j} - 10\mathbf{k}]$$
$$q_a - q_b = [1, \ -2\mathbf{j} + 2\mathbf{k}].$$

4.20.2 Norm of a Quaternion

Find the norm of q_a and q_b.

$$q_a = [2, \ -2\mathbf{i} + 3\mathbf{j} - 4\mathbf{k}], \qquad q_b = [1, \ -2\mathbf{i} + 5\mathbf{j} - 6\mathbf{k}].$$

Solution: Compute the square-root of the sum of the squares.

$$\|q_a\| = \sqrt{2^2 + (-2)^2 + 3^2 + (-4)^2} = \sqrt{33}$$
$$\|q_b\| = \sqrt{1^2 + (-2)^2 + 5^2 + (-6)^2} = \sqrt{66}.$$

4.20.3 Unit-Norm Form of a Quaternion

Convert q_a and q_b to their unit-norm form.

$$q_a = [2, \ -2\mathbf{i} + 3\mathbf{j} - 4\mathbf{k}], \qquad q_b = [1, \ -2\mathbf{i} + 5\mathbf{j} - 6\mathbf{k}].$$

Solution: Divide each quaternion by their norms calculated above.

$$\|q_a\| = \sqrt{33}$$
$$\|q_b\| = \sqrt{66}$$
$$q_a' = \tfrac{1}{\sqrt{33}} [2, \ -2\mathbf{i} + 3\mathbf{j} - 4\mathbf{k}]$$
$$q_b' = \tfrac{1}{\sqrt{66}} [1, \ -2\mathbf{i} + 5\mathbf{j} - 6\mathbf{k}].$$

4.20.4 Quaternion Product

Compute the product and reverse product of q_a and q_b.

$$q_a = [2, \ -2\mathbf{i} + 3\mathbf{j} - 4\mathbf{k}], \qquad q_b = [1, \ -2\mathbf{i} + 5\mathbf{j} - 6\mathbf{k}].$$

Solution: Compute $q_a q_b$ using $[s_a s_b - \mathbf{a} \cdot \mathbf{b}, \ s_a \mathbf{b} + s_b \mathbf{a} + \mathbf{a} \times \mathbf{b}]$. For the product $q_b q_a$, the cross-product vector is reversed.

$$q_a q_b = [2, \ -2\mathbf{i} + 3\mathbf{j} - 4\mathbf{k}][1, \ -2\mathbf{i} + 5\mathbf{j} - 6\mathbf{k}]$$

$$= [2 \times 1 - ((-2) \times (-2) + 3 \times 5 + (-4) \times (-6)),$$

$$2(-2\mathbf{i} + 5\mathbf{j} - 6\mathbf{k}) + 1(-2\mathbf{i} + 3\mathbf{j} - 4\mathbf{k})$$

$$+ (3 \times (-6) - (-4) \times 5)\mathbf{i} - ((-2) \times (-6) - (-4) \times (-2))\mathbf{j} + ((-2) \times 5 - 3 \times (-2))\mathbf{k}]$$

$$= [-41, \ -6\mathbf{i} + 13\mathbf{j} - 16\mathbf{k} + 2\mathbf{i} - 4\mathbf{j} - 4\mathbf{k}]$$

$$= [-41, \ -4\mathbf{i} + 9\mathbf{j} - 20\mathbf{k}]$$

$$q_b q_a = [1, \ -2\mathbf{i} + 5\mathbf{j} - 6\mathbf{k}][2 - 2\mathbf{i} + 3\mathbf{j} - 4\mathbf{k}]$$

$$= [1 \times 2 - ((-2) \times (-2) + 5 \times 3 + (-6) \times (-4)),$$

$$1(-2\mathbf{i} + 3\mathbf{j} - 4\mathbf{k}) + 2(-2\mathbf{i} + 5\mathbf{j} - 6\mathbf{k})$$

$$+ (5 \times (-4) - (-6) \times 3)\mathbf{i} - ((-2) \times (-4) - (-6) \times (-2))\mathbf{j} + ((-2) \times 3 - 5 \times (-2))\mathbf{k}]$$

$$= [-41, \ -6\mathbf{i} + 13\mathbf{j} - 16\mathbf{k} - 2\mathbf{i} + 4\mathbf{j} + 4\mathbf{k}]$$

$$= [-41, \ -8\mathbf{i} + 17\mathbf{j} - 12\mathbf{k}].$$

Note: The only thing that has changed in this computation is the sign of the cross-product axial vector.

4.20.5 Square of a Quaternion

Compute the square of q.

$$q = [2, \ -2\mathbf{i} + 3\mathbf{j} - 4\mathbf{k}].$$

Solution: Compute q^2 using $\left[s^2 - x^2 - y^2 - z^2, \ 2s(x\mathbf{i} + y\mathbf{j} + z\mathbf{k}) \right]$.

$$q^2 = [2, \ -2\mathbf{i} + 3\mathbf{j} - 4\mathbf{k}][2, \ -2\mathbf{i} + 3\mathbf{j} - 4\mathbf{k}]$$

$$= [2 \times 2 - ((-2) \times (-2) + 3 \times 3 + (-4) \times (-4)), +2 \times 2(-2\mathbf{i} + 3\mathbf{j} - 4\mathbf{k})]$$

$$= [-25, \ -8\mathbf{i} + 12\mathbf{j} - 16\mathbf{k}].$$

4.20.6 Inverse of a Quaternion

Compute the inverse of q.

$$q = [2, \ -2\mathbf{i} + 3\mathbf{j} - 4\mathbf{k}].$$

Solution: Compute the inverse using $q^{-1} = \overline{q}/\|q\|^2$.

$$\overline{q} = [2, \ +2\mathbf{i} - 3\mathbf{j} + 4\mathbf{k}]$$
$$\|q\|^2 = 2^2 + (-2)^2 + 3^2 + (-4)^2 = 33$$
$$q^{-1} = \tfrac{1}{33} [2, \ 2\mathbf{i} - 3\mathbf{j} + 4\mathbf{k}] .$$

References

1. Crowe MJ (1994) A history of vector analysis. Dover Publications, New York
2. Cheng H, Gupta KC (1989) An historical note on finite rotations. Trans ASME J Appl Mech 56(1):139–145
3. Altmann SL (1986) Rotations, quaternions and double groups. Dover Publications, New York, 2005. ISBN-13: 978-0-486-44518-2
4. Altmann SL (1989) Rodrigues, and the quaternion scandal. Math Mag 62(5):291–308
5. Altmann SL (1992) Icons and symmetries. Clarendon Press, Oxford
6. Altmann SL, Ortiz EL (eds) (2005) Mathematics and social Utopias in France: Olinde Rodrigues and his times, vol 28. History of mathematics. American Mathematical Society, Providence. ISBN-10: 0-8218-3860-1, ISBN-13: 978-0-8218-3860-0
7. Hamilton WR (1833) In: Conway AW, Synge JL (eds) The Mathematical Papers of Sir William Rowan Hamilton: Geometrical Optics, vol I; The Mathematical Papers of Sir William Rowan Hamilton: Dynamics, vol II; In: Conway AW, McDonnell AJ (eds); In: Halberstam H, Ingram RE (eds) The Mathematical Papers of Sir William Rowan Hamilton: Algebra, vol III. Cambridge University Press, Cambridge, 1931, 1940, 1967
8. Hamilton WR (1843). http://www-history.mcs.st-andrews.ac.uk/Mathematicians/Hamilton.html
9. Hamilton WR (1844) On quaternions: or a new system of imaginaries in algebra. Philos Mag ser 3 25: 1844
10. Hamilton WR (1853) Lectures on quaternions. Hodges & Smith, Dublin
11. Hamilton WR (1899–1901) Elements of quaternions. In: Jolly CJ (ed), vol 2, 2nd edn. Longmans, Green & Co., London
12. Tait PG (1867) Elementary treatise on quaternions, Cambridge University Press, Cambridge
13. Gauss CF (1819) Mutation des Raumes. Carl Friedrich Gauss Werke, Achter Band, König Gesell. Wissen, Göttingen, 1900, pp 357–361
14. Wilson EB (1901) Vector analysis. Yale University Press, New Haven
15. Feynman RP Symmetry and physical laws. Feynman lectures in physics, vol. 1

Chapter 5
Octonions

5.1 Introduction

Starting with a complex number $z = a + bi$, and extending this to a quaternion $q = [s + ai + bj + ck]$, it seems only natural to seek the existence of something similar with higher dimensions, which turns out to be an octonion, with eight elements. In this chapter I describe octonions, their algebraic properties and some worked examples. It is in no way a definitive exposition, but a gentle introduction to this obscure mathematical construct.

5.2 Background

A *division algebra* A possesses a multiplicative inverse $a^{-1} \in A$ for every non-zero element $a \in A$ such that

$$aa^{-1} = a^{-1}a = 1.$$

The multiplicative inverses are for a

$$\text{real number } x, \quad x^{-1} = 1/x, \qquad \mathbb{R}$$
$$\text{complex number } z = a + bi, \quad z^{-1} = \bar{z}/(a^2 + b^2), \quad \mathbb{C}$$
$$\text{quaternion } q = [s, v\hat{\mathbf{v}}], \quad q^{-1} = \bar{q}/(s^2 + v^2). \quad \mathbb{Q}$$

© Springer International Publishing AG, part of Springer Nature 2018
J. Vince, *Imaginary Mathematics for Computer Science*,
https://doi.org/10.1007/978-3-319-94637-5_5

The Euclidean norm of an element is a measure of its magnitude or length. Such norms are for a

$$\text{real number } x, \quad ||x|| = \sqrt{x^2}, \qquad \mathbb{R}$$
$$\text{complex number } z = a + bi, \quad ||z|| = \sqrt{a^2 + b^2}, \quad \mathbb{C}$$
$$\text{quaternion } q = [s, v\hat{\mathbf{v}}], \quad ||q|| = \sqrt{s^2 + v^2}. \quad \mathbb{Q}$$

A *normed division algebra* is one where

$$||ab|| = ||a|| \cdot ||b||.$$

For

$$\text{real numbers:} \quad ||ab|| = |a| \cdot |b|, \qquad \mathbb{R}$$
$$\text{complex numbers:} \quad ||z_1 z_2|| = ||z_1|| \cdot ||z_2||, \quad \mathbb{C}$$
$$\text{quaternions:} \quad ||q_a q_b|| = ||q_a|| \cdot ||q_b||. \quad \mathbb{Q}$$

When Hamilton's friend John Graves invented "octaves" in 1843, he wrote to Hamilton about his invention, showing that they were a normed division algebra requiring eight dimensions. Graves continued his research to look for a 16-dimensional version, which resulted in failure – this is because there are only four normed division algebras, as proved by Adolf Hurwitz in 1898 [1]. However, a 16-dimension algebra does exist: *sedenions* $\mathbb{S}$, but does not possess the properties of a normed division algebra.

The history of mathematics is littered with examples where two, or more, mathematicians come across the same idea simultaneously, and octonions are a good example. In 1845, the 24-year-old English mathematician Arthur Cayley published a paper on Jacobi's elliptic functions [2], which also included an appendix describing his discovery of an 8-dimensional normed division algebra. Unfortunately, for Graves, Cayley's paper was published first, and his discovery became known as "Cayley numbers". Today, they are known as octonions. Thus the four normed division algebras comprise: $\mathbb{R}$, $\mathbb{C}$, $\mathbb{Q}$ and $\mathbb{O}$.

John Baez describes the four normed division algebras as follows:

There are exactly four normed division algebras: the real numbers ($\mathbb{R}$), complex numbers ($\mathbb{C}$), quaternions ($\mathbb{H}$) and octonions ($\mathbb{O}$). The real numbers are the dependable breadwinner of the family, the complete ordered field we all rely on. The complex numbers are a slightly flashier but still respectable younger brother: not ordered, but algebraically complete. The quaternions, being noncommutative, are the eccentric cousin who is shunned at important family gatherings. But the octonions are the crazy old uncle nobody lets out of the attic: they are *nonassociative* [3].

So let's look closer at "the crazy old uncle in the attic"!

5.3 The Octonions

5.3.1 Notation

With complex numbers and quaternions defined respectively as

$$z = a + bi$$
$$q = s + ai + bj + ck$$

it follows that an octonion should be expressed as

$$x = s + ai + bj + ck + dl + eI + fJ + gK$$

where $\{s, a, b, c, d, e, f, g\} \in \mathbb{R}$ and $\{i, j, k, l, I, J, K\} \in \mathbb{I}$.

However, the following notation is also used, which is employed in geometric algebra:

$$x = x_0e_0 + x_1e_1 + x_2e_2 + x_3e_3 + x_4e_4 + x_5e_5 + x_6e_6 + x_7e_7$$

where $e_0 = 1$, $x_i \in \mathbb{R}$, $\{0 \leq i \leq 7\}$ $e_i \in \mathbb{I}$, $\{1 \leq i \leq 7\}$.

Although octonions obey the axioms of $\mathbb{R}$ for addition and subtraction, multiplication is non-commutative, like quaternions; but curiously they are non-associative. Table 5.1 shows the multiplication table for the octonion imaginary units i, j, k, l, I, J, K, and Table 5.2 shows an alternative multiplication table for the octonion imaginary units $e_1, e_2, \ldots, e_6, e_7$. Thus we still have

$$ij = k, \quad jk = i, \quad ki = j$$

Table 5.1 Multiplication table for the octonion imaginary units i, j, k, l, I, J, K

	i	j	k	l	I	J	K
i	-1	k	$-j$	I	$-l$	$-K$	J
j	$-k$	-1	i	J	K	$-l$	$-I$
k	j	$-i$	-1	K	$-J$	I	$-l$
l	$-I$	$-J$	$-K$	-1	i	j	k
I	l	$-K$	J	$-i$	-1	$-k$	j
J	K	l	$-I$	$-j$	k	-1	$-i$
K	$-J$	I	l	$-k$	$-j$	i	-1

Table 5.2 Multiplication table for the octonion imaginary units $e_1, e_2, \ldots, e_6, e_7$

		e_j						
	$e_i e_j$	e_1	e_2	e_3	e_4	e_5	e_6	e_7
	e_1	-1	e_3	$-e_2$	e_5	$-e_4$	$-e_7$	e_6
	e_2	$-e_3$	-1	e_1	e_6	e_7	$-e_4$	$-e_5$
e_i	e_3	e_2	$-e_1$	-1	e_7	$-e_6$	e_5	$-e_4$
	e_4	$-e_5$	$-e_6$	$-e_7$	-1	e_1	e_2	e_3
	e_5	e_4	$-e_7$	e_6	$-e_1$	-1	$-e_3$	e_2
	e_6	e_7	e_4	$-e_5$	$-e_2$	e_3	-1	$-e_1$
	e_7	$-e_6$	e_5	e_4	$-e_3$	$-e_2$	e_1	-1

together with

$$i^2 = j^2 = k^2 = l^2 = I^2 = J^2 = K^2 = -1.$$

To illustrate the non-associativity, consider the expression $l(JI)$. We evaluate (JI) first, which gives k, then premultiply by l, which results in $-K$. Changing the expression to $(lJ)I$, where evaluating (lJ) gives j, which post-multiplied by I, results in K. Thus great care must be taken when evaluating multiple products.

5.3.2 Cayley–Dickson Construction

The Cayley–Dickson construction, named after Arthur Cayley and the American mathematician Leonard Dickson (1874–1954), generalises Hamilton's substitution of an ordered pair of real numbers for a complex number, to quaternions and octonions. The construction describes one algebra as a pair of elements from an algebra of a lower dimension. Thus, an octonion is an ordered pair of quaternions, which are ordered pairs of complex numbers, which, in turn, are ordered pairs of real numbers.

For example, a complex number z is defined by a pair of real numbers as follows:

$$z = (a, b) = a + bi$$

and the product of two complex numbers is defined by:

ordered pair notation	complex notation
$z_1 = (a,\ b)$	$z_1 = a + bi$
$z_2 = (c,\ d)$	$z_2 = c + di$
$(a,\ b)(c,\ d) = (ac - bd,\ ad + bc)$	$z_1 z_2 = ac - bd + (ad + bc)i.$

The conjugate of z: $\bar{z} = a - bi$, becomes the ordered pair $(a, -b)$, which makes

$$z\bar{z} = (a, \ b)(a, \ -b) = (a^2 + b^2, \ 0)$$

and is used as the basis for the Euclidean norm of z:

$$\|z\| = |z| = \sqrt{z\bar{z}}.$$

Next, a quaternion q is defined by a pair of complex numbers as follows:

$$a = a_1 + a_2 i$$
$$b = b_1 + b_2 i$$
$$q = (a, \ b) = a + bj$$
$$= a_1 + a_2 i + (b_1 + b_2 i)j$$
$$= a_1 + a_2 i + b_1 j + b_2 k$$

which is a quaternion. Note that in the above, $ij = k$.

The product of two quaternions is defined by two pairs of complex numbers

$$a = a_1 + a_2 i, \quad b = b_1 + b_2 i$$
$$c = c_1 + c_2 i, \quad d = d_1 + d_2 i$$
$$(a, b)(c, d) = (a_1 + a_2 i + b_1 j + b_2 k)(c_1 + c_2 i + d_1 j + d_2 k)$$
$$= (a_1 c_1 - a_2 c_2 - b_1 d_1 - b_2 d_2)$$
$$+ (a_1 c_2 + b_1 d_2 + a_2 c_1 - b_2 d_1)i$$
$$+ (a_1 d_1 - a_2 d_2 + b_1 c_1 + b_2 c_2)j$$
$$+ (a_1 d_2 + a_2 d_1 - b_1 c_2 + b_2 c_1)k$$
$$= (a_1 c_1 - a_2 c_2) + (a_1 c_2 + a_2 c_1)i$$
$$- [(b_1 d_1 + b_2 d_2) + (-b_1 d_2 + b_2 d_1)i]$$
$$+ [(a_1 d_1 - a_2 d_2) + (a_1 d_2 + a_2 d_1)i]j$$
$$+ [(b_1 c_1 + b_2 c_2) + (-b_1 c_2 + b_2 c_1)i]j$$
$$= (ac - b\bar{d}) + (ad + b\bar{c})j$$
$$= (ac - b\bar{d}, \ ad + b\bar{c}).$$

Thus the product of two quaternions represented by two ordered pairs of complex numbers is:

$$(a, \ b)(c, \ d) = (ac - b\bar{d}, \ ad + b\bar{c}).$$

The only difference between this and the expression for the product of two complex numbers, is the introduction of the conjugate operation, which has no effect on real numbers.

The Cayley–Dickson construction shows that this relationship holds for octonions, such that defining an octonion from two pairs of quaternions $(a,\ b)$ and $(c,\ d)$, their product is

$$(a,\ b)(c,\ d) = (ac - b\overline{d},\ ad + b\overline{c}).$$

5.4   Octonion Algebra

Starting with the definition for an octonion x as

$$x = (x_0 e_0,\ x_1 e_1, \ldots, x_7 e_7)\ = x_0 e_0 + \sum_{i=1}^{7} x_i e_i$$

where $e_0 = 1,\quad x_i \in \mathbb{R}\ \{0 \le i \le 7\},\quad e_i \in \mathbb{I}\ \{1 \le i \le 7\}.$

An octonion comprises a real scalar part x_0 and a vector part $\sum_{i=1}^{7} x_i e_i$.

5.4.1   Octonion Addition and Subtraction

Two octonions are added or subtracted like complex numbers and quaternions, simply by resolving pairs of terms. For example, two octonions are added and subtracted as follows:

$$x = \sum_{i=0}^{7} x_i e_i,\quad y = \sum_{i=0}^{7} y_i e_i$$

$$x \pm y = \sum_{i=0}^{7} (x_i \pm y_i) e_i.$$

$$a = 3 + 2i + 4j - 5k + 6I + 2K$$
$$b = 2 + 2i + 2j - 3k + 2l + 3K$$
$$a + b = 5 + 4i + 6j - 8k + 2l + 6I + 5K$$
$$a - b = 1 + 2j - 2k - 2l + 6I - K.$$

5.4.2 Octonion Multiplication

In order to simplify the algebra, I will employ two octonions with only three terms:

$$a = 2 + 3j + 4I$$
$$b = 2i + k + 3K$$
$$ab = (2 + 3j + 4I)(2i + k + 3K)$$
$$= 2(2i + k + 3K) + 3j(2i + k + 3K) + 4I(2i + k + 3K)$$
$$= 4i + 2k + 6K - 6k + 3i - 9I + 8l + 4J + 12j$$
$$= 7i + 12j - 4k - 9I + 4J + 6K.$$

Let's reverse the product sequence:

$$a = 2 + 3j + 4I$$
$$b = 2i + k + 3K$$
$$ba = (2i + k + 3K)(2 + 3j + 4I)$$
$$= 2i(2 + 3j + 4I) + k(2 + 3j + 4I) + 3K(2 + 3j + 4I)$$
$$= 4i + 6k - 8l + 2k - 3i - 4J + 6K + 9I - 12j$$
$$= i - 12j + 8k - 8l + 9I - 4J + 6K.$$

Which confirms that octonions do not commute.

5.4.3 Octonion Conjugate

As with complex numbers and quaternions, the conjugate operation reverses the sign of the imaginary part:

$$\overline{x} = x_0 - \sum_{i=1}^{7} x_i e_i.$$

For example,

$$x = -12 + 3j + 6I - 4K$$
$$\overline{x} = -12 - 3j - 6I + 4K.$$

The real and imaginary parts of an octonion are isolated algebraically using:

$$x_0 = \tfrac{1}{2}(x + \overline{x}), \qquad \sum_{i=1}^{7} x_i e_i = \tfrac{1}{2}(x - \overline{x}).$$

Conjugating the product of two octonions is defined as

$$(\overline{xy}) = \overline{y}\,\overline{x}.$$

The product $x\overline{x}$ or $\overline{x}x$ always results in a nonnegative real value:

$$x\overline{x} = \sum_{i=0}^{7} x_i^2.$$

For example,

$$
\begin{aligned}
x &= 2 + 3j + 4J \\
\overline{x} &= 2 - 3j - 4J \\
x\overline{x} &= (2 + 3j + 4J)(2 - 3j - 4J) \\
&= 2(2 - 3j - 4J) + 3j(2 - 3j - 4J) + 4J(2 - 3j - 4J) \\
&= 4 - 6j - 8J + 6j + 9 + 12l + 8J - 12l + 16 \\
&= 29.
\end{aligned}
$$

5.4.4 Norm of an Octonion

The norm of an octonion, which is also the Euclidean norm on $\mathbb{R}^8$, is defined as

$$\|x\| = \sqrt{x\overline{x}}.$$

For example,

$$x = 3 + 2j + 3I + 4K$$

$$x\overline{x} = \sum_{i=0}^{7} x_i^2 = 3^2 + 2^2 + 3^2 + 4^2 = 38$$

$$\|x\| = \sqrt{38} \approx 6.16.$$

5.4.5   *Inverse of an Octonion*

The inverse of a non-zero octonion is given by

$$x^{-1} = \frac{\overline{x}}{\|x\|^2}.$$

For example,

$$x = 2 + 3j + 4J - 2K$$
$$\overline{x} = 2 - 3j - 4J + 2K$$
$$\|x\|^2 = x\overline{x} = 2^2 + 3^2 + 4^2 + 2^2 = 33$$
$$x^{-1} = \frac{\overline{x}}{\|x\|^2} = \tfrac{1}{33}(2 - 3j - 4J + 2K)$$

One can see that the product $xx^{-1} = 1$.

5.5   Summary of Operations

Octonion

$$x = s + ai + bj + ck + dl + eI + fJ + gK$$
$$\text{where } \{s, a, b, c, d, e, f, g\} \in \mathbb{R} \quad \text{and} \quad \{i, j, k, l, I, J, K\} \in \mathbb{I}$$

or

$$x = x_0 e_0 + x_1 e_1 + x_2 e_2 + x_3 e_3 + x_4 e_4 + x_5 e_5 + x_6 e_6 + x_7 e_7$$
$$\text{where } e_0 = 1, \quad x_i \in \mathbb{R} \ \{0 \leq i \leq 7\}, \quad e_i \in \mathbb{I} \ \{1 \leq i \leq 7\}.$$

Adding and Subtracting

$$x = \sum_{i=0}^{7} x_i e_i, \quad y = \sum_{i=0}^{7} y_i e_i$$

$$x \pm y = \sum_{i=0}^{7} (x_i \pm y_i) e_i.$$

Conjugate

$$\overline{x} = x_0 - \sum_{i=1}^{7} x_i e_i.$$

Norm

$$\|x\| = \sqrt{x\overline{x}}.$$

Inverse

$$x^{-1} = \frac{\overline{x}}{\|x\|^2}.$$

5.6   Worked Examples

5.6.1   Adding and Subtracting Octonions

Given two octonions x and y, calculate $x + y$ and $x - y$.
Solution: Add and subtract the respective elements.

$$x = 2 + 3i + 4j + 5k + 6l - 7I + 8J - 9K$$
$$y = 1 + 2i + 2j - 4k + 4I + 7K$$
$$x + y = 3 + 5i + 6j + k + 6l - 3I + 8J - 2K$$
$$x - y = 1 + i + 2j + 9k + 6l - 11I + 8J - 16K.$$

5.6.2   Multiplying Two Octonions

Given two octonions x and y, calculate their product xy.
Solution: Expand algebraically the product terms using Table 5.1.

$$x = 2 + 3i + 4J + 2K$$
$$y = 3 + 2j + 2l + 2K$$
$$xy = 2(3 + 2j + 2l + 2K) + 3i\,(3 + 2j + 2l + 2K)$$
$$\quad + 4J\,(3 + 2j + 2l + 2K) + 2K\,(3 + 2j + 2l + 2K)$$
$$= 6 + 4j + 4l + 4K + 9i + 6k + 6I + 6J$$
$$\quad + 12J + 8l - 8j - 8i + 6K + 4l - 4k - 4$$
$$= 2 + i - 4j + 2k + 16l + 6I + 18J + 10K.$$

5.6.3   Conjugate of an Octonion

State the conjugate of octonion $x = 12 + 3i - 6j + 7k + 2l - 6I + 4J + 3K$.
Solution: Reverse the signs of the imaginary elements.

$$\bar{x} = 12 - 3i + 6j - 7k - 2l + 6I - 4J - 3K.$$

5.6.4   Norm of an Octonion

Calculate the Euclidean norm of $x = 2 + 3i + 3J + 4K$.
Solution: Compute the square-root of the sum of the squares of the scalar terms.

$$\|x\| = \sqrt{2^2 + 3^2 + 3^2 + 4^2} = \sqrt{38} \approx 6.16.$$

5.6.5   Inverse of an Octonion

Calculate the inverse of octonion $x = 2 + 3i + 3J + 4K$.
Solution: Divide the conjugate of x by its norm squared, calculated above.

$$\bar{x} = 2 - 3i - 3J - 4K$$
$$\|x\|^2 = 38$$
$$x^{-1} = \tfrac{1}{38}(2 - 3i - 3J - 4K).$$

References

1. Hurwitz A (1898) ber die Composition der quadratischen Formen von beliebig vielen Variabeln. Goett. Nachr. 309316
2. Cayley A (1845) On Jacobi's elliptic functions, in reply to the Rev. B. Brownwin; and on quaternions. Philos. Mag. 26(1845):208–211
3. Baez JC http://www.math.ucr.edu/home/baez/octonions/node1.html

Chapter 6
Geometric Algebra

6.1 Introduction

This can only be a brief introduction to geometric algebra as the subject really demands an entire book. Those readers who wish to pursue the subject further should consult the author's books [1, 2].

Complex numbers, quaternions and octonions, explicitly define objects such as i^2, j^2 and k^2 that equal -1, from which a complex algebra is constructed. Geometric algebra, on the other hand, reverses the sequence by developing an algebra for geometric analysis, and reveals that the underlying algebraic constructs are imaginary. This chapter begins by examining some trigonometric foundations associated with line segments, defines two geometric products, and then reveals the hidden imaginary properties.

A challenge for computer science is to develop software that undertakes the symbolic manipulation that is revealed in this chapter.

6.2 Background

Before Hamilton discovered quaternions, Hermann Grassmann had developed his own theory of vectors, but was unable to influence the tide of opinion, even though the first application of his notation was employed in a 200-page essay on the theory of tides *Theorie der Ebbe und Flut* [3].

By the early 20th century, vector analysis had been determined by Josiah Willard Gibbs, who was not an admirer of Hamilton's quaternions. Gibbs recognised that a pure quaternion could be interpreted as a vector without any imaginary connotation, and could form the basis of a vectorial system. Two products for vectors emerged: the dot (scalar) product and the cross (vector) product, which could be combined to form the scalar triple product and the vector triple product. At last, vector analysis

© Springer International Publishing AG, part of Springer Nature 2018
J. Vince, *Imaginary Mathematics for Computer Science*,
https://doi.org/10.1007/978-3-319-94637-5_6

had been defined and was understood. But mathematicians were unaware that they had walked up a mathematical cul de sac!

Every student of mathematics knows that the cross product has no meaning in 2D, behaves immaculately in 3D, but is ambiguous in higher dimensional spaces. So because of its inherent fussiness, it is not an important mathematical product after all, in spite of its usefulness in resolving 3D geometric problems.

Geometric algebra proposes an alternative vectorial framework where lines, areas, volumes and hyper-volumes are recognised as structures with magnitude and orientation. Oriented lines are represented by vectors, oriented areas by bivectors and oriented volumes by trivectors. Higher dimensional objects are also permitted. At the heart of geometric algebra is the geometric product, which is defined as the sum of the inner and outer products. The inner product is related to the scalar product, and the outer product is related to the cross product. What is so flexible about this approach is that all sorts of products are permitted such as (line × line), (line × area), (area × area), (line × volume), (area × volume), (volume × volume), etc. Furthermore, the cross product has its alias within the algebra as do quaternions; and on top of these powerful features one can add, subtract and even divide such oriented objects.

You are probably wondering how it is possible that such a useful algebra has lain dormant for so many years? Well, through the endeavours of the English mathematician William Kingdon Clifford (1845–1879), and the American theoretical physicist David Orlin Hestenes (1933–), we now have a geometric calculus that is being embraced by the physics community through the work of Anthony Lasenby, Joan Lasenby and Chris Doran. In spite of geometric algebra's struggle to surface, today it does exist and is relatively easy to understand, and I will reveal its axioms and structures in the following chapter.

6.3 Symmetric and Anti-symmetric Functions

It is useful to classify functions into two categories: *symmetric* (*even*) and *anti-symmetric* (*odd*) functions. For example, given two symmetric functions $f(x)$ and $f(x, y)$:

$$f(-x) = f(x)$$

and

$$f(y, x) = f(x, y)$$

an example being $\cos x$ where $\cos(-x) = \cos x$. Figure 6.1 illustrates how the cosine function is reflected about the origin. However, if the functions are anti-symmetric:

Fig. 6.1 The graph of the symmetric cosine function

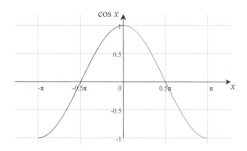

Fig. 6.2 The graph of the anti-symmetric sine function

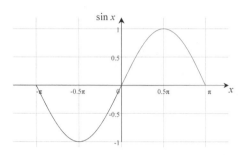

$$f(-x) = -f(x)$$

and

$$f(y, x) = -f(x, y)$$

an example being $\sin x$ where $\sin(-x) = -\sin x$. Figure 6.2 illustrates how the sine function is reflected about the origin.

6.4 Trigonometric Foundations

Figure 6.3 shows two line segments a and b with coordinates (a_1, a_2), (b_1, b_2) respectively. The lines are separated by an angle θ, and I will compute the expressions $ab\cos\theta$ and $ab\sin\theta$, as these play an important role in geometric algebra.

Using the trigonometric identities

$$\sin(\theta + \phi) = \sin\theta\cos\phi + \cos\theta\sin\phi \tag{6.1}$$
$$\cos(\theta + \phi) = \cos\theta\cos\phi - \sin\theta\sin\phi \tag{6.2}$$

and the following observations

Fig. 6.3 Two line segments
a and *b* separated by $+\theta$

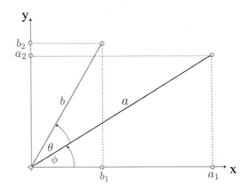

$$\cos\phi = \frac{a_1}{a}, \quad \sin\phi = \frac{a_2}{a}, \quad \cos(\theta + \phi) = \frac{b_1}{b}, \quad \sin(\theta + \phi) = \frac{b_2}{b}$$

I can rewrite (6.1) and (6.2) as

$$\frac{b_2}{b} = \frac{a_1}{a}\sin\theta + \frac{a_2}{a}\cos\theta \tag{6.3}$$

$$\frac{b_1}{b} = \frac{a_1}{a}\cos\theta - \frac{a_2}{a}\sin\theta. \tag{6.4}$$

To isolate $\cos\theta$, multiply (6.3) by a_2 and (6.4) by a_1:

$$\frac{a_2 b_2}{b} = \frac{a_1 a_2}{a}\sin\theta + \frac{a_2^2}{a}\cos\theta \tag{6.5}$$

$$\frac{a_1 b_1}{b} = \frac{a_1^2}{a}\cos\theta - \frac{a_1 a_2}{a}\sin\theta. \tag{6.6}$$

Adding (6.5) and (6.6):

$$\frac{a_1 b_1 + a_2 b_2}{b} = \frac{a_1^2 + a_2^2}{a}\cos\theta = a\cos\theta$$

therefore,

$$ab\cos\theta = a_1 b_1 + a_2 b_2.$$

To isolate $\sin\theta$, multiply (6.3) by a_1 and (6.4) by a_2

$$\frac{a_1 b_2}{b} = \frac{a_1^2}{a}\sin\theta + \frac{a_1 a_2}{a}\cos\theta \tag{6.7}$$

$$\frac{a_2 b_1}{b} = \frac{a_1 a_2}{a}\cos\theta - \frac{a_2^2}{a}\sin\theta \tag{6.8}$$

Fig. 6.4 Two line segments
a and b separated by $-\theta$

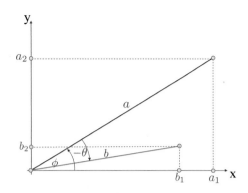

Subtracting (6.8) from (6.7):

$$\frac{a_1 b_2 - a_2 b_1}{b} = \frac{a_1^2 + a_2^2}{a}\sin\theta = a\sin\theta$$

therefore,

$$ab\sin\theta = a_1 b_2 - a_2 b_1.$$

If we form the product of b's projection on a with a, we get $ab\cos\theta$ which we have shown equals $a_1 b_1 + a_2 b_2$. Similarly, if we form the product $ab\sin\theta$ we compute the area of the parallelogram formed by sweeping a along b, which equals $a_1 b_2 - a_2 b_1$. What is noteworthy, is that the product $ab\cos\theta$ is independent of the sign of the angle θ, whereas the product $ab\sin\theta$ is sensitive to the sign of θ. Consequently, if we construct the lines a and b such that b is rotated $-\theta$ relative to a as shown in Fig. 6.4, $ab\cos\theta = a_1 b_1 + a_2 b_2$, but $ab\sin\theta = -(a_1 b_2 - a_2 b_1)$. The anti-symmetric nature of the sine function reverses the sign of the area.

Having shown that area is a signed quantity just by using trigonometric identities, let's explore how vector algebra responds to this idea.

6.5 Vectorial Foundations

The algebraic product of two 2D vectors $\mathbf{a}$ and $\mathbf{b}$ is

$$\mathbf{a} = a_1 \mathbf{i} + a_2 \mathbf{j}$$
$$\mathbf{b} = b_1 \mathbf{i} + b_2 \mathbf{j}$$
$$\mathbf{ab} = a_1 b_1 \mathbf{i}^2 + a_2 b_2 \mathbf{j}^2 + a_1 b_2 \mathbf{ij} + a_2 b_1 \mathbf{ji} \qquad (6.9)$$

and it is clear that $a_1b_1\mathbf{i}^2 + a_2b_2\mathbf{j}^2$ has something to do with $ab\cos\theta$, and $a_1b_2\mathbf{ij} + a_2b_1\mathbf{ji}$ has something to do with $ab\sin\theta$. The product $\mathbf{ab}$ creates the terms $\mathbf{i}^2$, $\mathbf{j}^2$, $\mathbf{ij}$ and $\mathbf{ji}$, which are resolved as follows.

6.6 Inner and Outer Products

Let's begin with the products $\mathbf{ij}$ and $\mathbf{ji}$ in (6.9) and assume that they anti-commute: $\mathbf{ji} = -\mathbf{ij}$. Therefore,

$$\mathbf{ab} = a_1b_1\mathbf{i}^2 + a_2b_2\mathbf{j}^2 + (a_1b_2 - a_2b_1)\mathbf{ij} \tag{6.10}$$

and if we reverse the product to $\mathbf{ba}$ we obtain

$$\mathbf{ba} = a_1b_1\mathbf{i}^2 + a_2b_2\mathbf{j}^2 - (a_1b_2 - a_2b_1)\mathbf{ij}. \tag{6.11}$$

From (6.10) and (6.11) we see that the product of two vectors contains a symmetric component

$$a_1b_1\mathbf{i}^2 + a_2b_2\mathbf{j}^2$$

and an anti-symmetric component

$$(a_1b_2 - a_2b_1)\mathbf{ij}.$$

It is interesting to observe that the symmetric component has $0°$ between its vector pairs ($\mathbf{i}^2$ and $\mathbf{j}^2$), whereas the anti-symmetric component has $90°$ between its vector pairs ($\mathbf{i}$ and $\mathbf{j}$). Therefore, the sine and cosine functions play a natural role in our rules. What we are looking for are two functions that, when given our vectors $\mathbf{a}$ and $\mathbf{b}$, one function returns the symmetric component and the other returns the anti-symmetric component. We call these the *inner* and *outer* functions respectively.

It should be clear that if the inner function includes the cosine of the angle between the two vectors it will reject the anti-symmetric component and return the symmetric element. Similarly, if the outer function includes the sine of the angle between the vectors, the symmetric component is rejected, and returns the anti-symmetric element.

Let's declare the inner function as the *inner product*

$$\mathbf{a} \cdot \mathbf{b} = |\mathbf{a}||\mathbf{b}|\cos\theta \tag{6.12}$$

then

$$\mathbf{a} \cdot \mathbf{b} = (a_1\mathbf{i} + a_2\mathbf{j}) \cdot (b_1\mathbf{i} + b_2\mathbf{j})$$
$$= a_1b_1\mathbf{i} \cdot \mathbf{i} + a_1b_2\mathbf{i} \cdot \mathbf{j} + a_2b_1\mathbf{j} \cdot \mathbf{i} + a_2b_2\mathbf{j} \cdot \mathbf{j}$$
$$= a_1b_1 + a_2b_2$$

which is perfect!

Next, let's declare the outer function as the *outer product* using the wedge "∧" symbol; which is why it is also called the *wedge product*:

$$\mathbf{a} \wedge \mathbf{b} = |\mathbf{a}||\mathbf{b}| \sin\theta \, \mathbf{i} \wedge \mathbf{j}. \qquad (6.13)$$

Note that product includes a strange $\mathbf{i} \wedge \mathbf{j}$ term. This is included as we just can't ignore the $\mathbf{ij}$ term in the anti-symmetric component:

$$\mathbf{a} \wedge \mathbf{b} = (a_1\mathbf{i} + a_2\mathbf{j}) \wedge (b_1\mathbf{i} + b_2\mathbf{j})$$
$$= a_1b_1\mathbf{i} \wedge \mathbf{i} + a_1b_2\mathbf{i} \wedge \mathbf{j} + a_2b_1\mathbf{j} \wedge \mathbf{i} + a_2b_2\mathbf{j} \wedge \mathbf{j}$$
$$= (a_1b_2 - a_2b_1)\mathbf{i} \wedge \mathbf{j}$$

which permits us to write

$$\mathbf{ab} = \mathbf{a} \cdot \mathbf{b} + \mathbf{a} \wedge \mathbf{b} \qquad (6.14)$$
$$\mathbf{ab} = |\mathbf{a}||\mathbf{b}| \cos\theta + |\mathbf{a}||\mathbf{b}| \sin\theta \, \mathbf{i} \wedge \mathbf{j}. \qquad (6.15)$$

6.7 The Geometric Product in 2D

Clifford named the sum of the two products the *geometric product*, which means that (6.14) reads: The geometric product $\mathbf{ab}$ is the sum of the inner product "$\mathbf{a}$ dot $\mathbf{b}$" and the outer product "$\mathbf{a}$ wedge $\mathbf{b}$". Remember that all this assumes that $\mathbf{ji} = -\mathbf{ij}$ which seems a reasonable assumption.

Given the definition of the geometric product, let's evaluate $\mathbf{i}^2$

$$\mathbf{ii} = \mathbf{i} \cdot \mathbf{i} + \mathbf{i} \wedge \mathbf{i}.$$

Using the definition for the inner product (6.12) we have

$$\mathbf{i} \cdot \mathbf{i} = 1 \times 1 \times \cos 0° = 1$$

whereas, using the definition of the outer product (6.13) we have

$$\mathbf{i} \wedge \mathbf{i} = 1 \times 1 \times \sin 0° \ \mathbf{i} \wedge \mathbf{i} = 0.$$

Thus $\mathbf{i}^2 = 1$ and $\mathbf{j}^2 = 1$, and $\mathbf{aa} = |\mathbf{a}|^2$:

$$\mathbf{aa} = \mathbf{a} \cdot \mathbf{a} + \mathbf{a} \wedge \mathbf{a}$$
$$= |\mathbf{a}||\mathbf{a}| \cos 0° + |\mathbf{a}||\mathbf{a}| \sin 0° \mathbf{i} \wedge \mathbf{j}$$
$$\mathbf{aa} = |\mathbf{a}|^2.$$

Now let's evaluate $\mathbf{ij}$:

$$\mathbf{ij} = \mathbf{i} \cdot \mathbf{j} + \mathbf{i} \wedge \mathbf{j}.$$

Using the definition for the inner product (6.12) we have

$$\mathbf{i} \cdot \mathbf{j} = 1 \times 1 \times \cos 90° = 0$$

whereas using the definition of the outer product (6.13) we have

$$\mathbf{i} \wedge \mathbf{j} = 1 \times 1 \times \sin 90° \ \mathbf{i} \wedge \mathbf{j} = \mathbf{i} \wedge \mathbf{j}.$$

Thus $\mathbf{ij} = \mathbf{i} \wedge \mathbf{j}$. But what is $\mathbf{i} \wedge \mathbf{j}$? Well, it is a new object called a *bivector*, and defines the orientation of the plane containing $\mathbf{i}$ and $\mathbf{j}$.

As the order of the vectors is from $\mathbf{i}$ to $\mathbf{j}$, the angle is $+90°$ and $\sin(+90)° = 1$. Whereas, if the order is from $\mathbf{j}$ to $\mathbf{i}$ the angle is $-90°$ and $\sin(-90°) = -1$. Consequently,

$$\mathbf{ji} = \mathbf{j} \cdot \mathbf{i} + \mathbf{j} \wedge \mathbf{i}$$
$$= 0 + 1 \times 1 \times \sin(-90°)\mathbf{i} \wedge \mathbf{j}$$
$$\mathbf{ji} = -\mathbf{i} \wedge \mathbf{j}.$$

Thus the bivector $\mathbf{i} \wedge \mathbf{j}$ defines the orientation of a surface as anti-clockwise, whilst the bivector $\mathbf{j} \wedge \mathbf{i}$ defines the orientation as clockwise. These ideas are shown in Fig. 6.5.

Fig. 6.5 An anti-clockwise and clockwise bivector

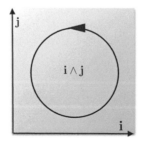 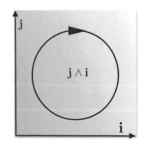

The inner product (6.12) is our old friend the dot product, and does not need explaining. However, the outer product (6.13) does require some further explanation.

The equation

$$ab = 9 + 12\mathbf{i} \wedge \mathbf{j}$$

simply means that the geometric product of two vectors $\mathbf{a}$ and $\mathbf{b}$ creates a scalar, inner product of 9, and an outer product of 12 on the $\mathbf{ij}$-plane.

For example, given

$$\mathbf{a} = 3\mathbf{i}$$
$$\mathbf{b} = 3\mathbf{i} + 4\mathbf{j}$$

then

$$\mathbf{ab} = 3\mathbf{i} \cdot (3\mathbf{i} + 4\mathbf{j}) + 3\mathbf{i} \wedge (3\mathbf{i} + 4\mathbf{j})$$
$$= 9 + 9\mathbf{i} \wedge \mathbf{i} + 12\mathbf{i} \wedge \mathbf{j}$$
$$\mathbf{ab} = 9 + 12\mathbf{i} \wedge \mathbf{j}.$$

The 9 represents $|\mathbf{a}||\mathbf{b}|\cos\theta$, whereas the 12 represents an area $|\mathbf{a}||\mathbf{b}|\sin\theta$ on the $\mathbf{ij}$-plane. The angle between the two vectors θ is given by

$$\theta = \cos^{-1}(3/5).$$

However, reversing the product, we obtain

$$\mathbf{ba} = (3\mathbf{i} + 4\mathbf{j}) \cdot 3\mathbf{i} + (3\mathbf{i} + 4\mathbf{j}) \wedge 3\mathbf{i}$$
$$= 9 + 9\mathbf{i} \wedge \mathbf{i} + 12\mathbf{j} \wedge \mathbf{i}$$
$$\mathbf{ab} = 9 - 12\mathbf{i} \wedge \mathbf{j}.$$

The sign of the outer (wedge) product has flipped to reflect the new orientation of the vectors relative to the accepted orientation of the basis bivectors.

So the geometric product combines the scalar and wedge products into a single product, where the scalar product is the symmetric component and the wedge product is the anti-symmetric component. Now let's see how these products behave in 3D.

6.8 The Geometric Product in 3D

Before we consider the geometric product in 3D we need to introduce some new notation, which will simplify future algebraic expressions. Rather than use $\mathbf{i}, \mathbf{j}$ and $\mathbf{k}$ to represent the unit basis vectors let's employ $\mathbf{e}_1, \mathbf{e}_2$ and $\mathbf{e}_3$ respectively. This means that (6.15) can be written

$$\mathbf{ab} = |\mathbf{a}||\mathbf{b}| \cos \theta + |\mathbf{a}||\mathbf{b}| \sin \theta \ \mathbf{e}_1 \wedge \mathbf{e}_2.$$

We begin with two 3D vectors:

$$\mathbf{a} = a_1 \mathbf{e}_1 + a_2 \mathbf{e}_2 + a_3 \mathbf{e}_3$$
$$\mathbf{b} = b_1 \mathbf{e}_1 + b_2 \mathbf{e}_2 + b_3 \mathbf{e}_3$$

therefore, their inner product is

$$\mathbf{a} \cdot \mathbf{b} = (a_1 \mathbf{e}_1 + a_2 \mathbf{e}_2 + a_3 \mathbf{e}_3) \cdot (b_1 \mathbf{e}_1 + b_2 \mathbf{e}_2 + b_3 \mathbf{e}_3)$$
$$= a_1 b_1 + a_2 b_2 + a_3 b_3$$

and their outer product is

$$\mathbf{a} \wedge \mathbf{b} = (a_1 \mathbf{e}_1 + a_2 \mathbf{e}_2 + a_3 \mathbf{e}_3) \wedge (b_1 \mathbf{e}_1 + b_2 \mathbf{e}_2 + b_3 \mathbf{e}_3)$$
$$= a_1 b_2 \mathbf{e}_1 \wedge \mathbf{e}_2 + a_1 b_3 \mathbf{e}_1 \wedge \mathbf{e}_3 + a_2 b_1 \mathbf{e}_2 \wedge \mathbf{e}_1 + a_2 b_3 \mathbf{e}_2 \wedge \mathbf{e}_3$$
$$+ a_3 b_1 \mathbf{e}_3 \wedge \mathbf{e}_1 + a_3 b_2 \mathbf{e}_3 \wedge \mathbf{e}_2$$

$$\mathbf{a} \wedge \mathbf{b} = (a_1 b_2 - a_2 b_1)\mathbf{e}_1 \wedge \mathbf{e}_2 + (a_2 b_3 - a_3 b_2)\mathbf{e}_2 \wedge \mathbf{e}_3 + (a_3 b_1 - a_1 b_3)\mathbf{e}_3 \wedge \mathbf{e}_1.$$
$$(6.16)$$

This time we have three unit-basis bivectors: $\mathbf{e}_1 \wedge \mathbf{e}_2, \mathbf{e}_2 \wedge \mathbf{e}_3, \mathbf{e}_3 \wedge \mathbf{e}_1$, and three associated scalar multipliers: $(a_1 b_2 - a_2 b_1), (a_2 b_3 - a_3 b_2), (a_3 b_1 - a_1 b_3)$ respectively.

Thus the geometric product of two 3D vectors remains $\mathbf{ab} = \mathbf{a} \cdot \mathbf{b} + \mathbf{a} \wedge \mathbf{b}$:

$$\mathbf{ab} = (a_1 b_1 + a_2 b_2 + a_3 b_3)$$
$$+ (a_1 b_2 - a_2 b_1)\mathbf{e}_1 \wedge \mathbf{e}_2 + (a_2 b_3 - a_3 b_2)\mathbf{e}_2 \wedge \mathbf{e}_3 + (a_3 b_1 - a_1 b_3)\mathbf{e}_3 \wedge \mathbf{e}_1.$$

Continuing with the idea described in the previous section, the three bivectors represent the three planes containing the respective vectors as shown in Fig. 6.6, and the scalar multipliers are projections of the area of the vector parallelogram onto the three bivectors as shown in Fig. 6.7. The orientation of the vectors $\mathbf{a}$ and $\mathbf{b}$ determine whether the projected areas are positive or negative.

You may think that (6.16) looks familiar. In fact, it looks very similar to the cross product $\mathbf{a} \times \mathbf{b}$:

$$\mathbf{a} \times \mathbf{b} = (a_1 b_2 - a_2 b_1)\mathbf{e}_3 + (a_2 b_3 - a_3 b_2)\mathbf{e}_1 + (a_3 b_1 - a_1 b_3)\mathbf{e}_2. \quad (6.17)$$

This similarity is no accident. For when Hamilton invented quaternions, he did not recognise the possibility of bivectors, and invented some rules, which eventually

Fig. 6.6 The 3D bivectors

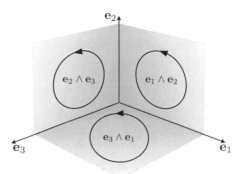

Fig. 6.7 The projections on
the three bivectors

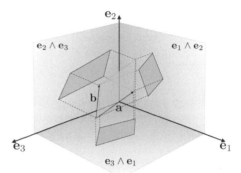

became the cross product! Later in this chapter we discover that quaternions are
really bivectors in disguise.

We can see that a simple relationship exists between (6.16) and (6.17):

$$\mathbf{e}_1 \wedge \mathbf{e}_2 \text{ and } \mathbf{e}_3$$
$$\mathbf{e}_2 \wedge \mathbf{e}_3 \text{ and } \mathbf{e}_1$$
$$\mathbf{e}_3 \wedge \mathbf{e}_1 \text{ and } \mathbf{e}_2$$

the wedge product bivectors are perpendicular to the vector components of the cross
product. So the wedge product is just another way of representing the cross product.
However, the wedge product introduces a very important bonus: it works in space of
any dimension, whereas, the cross product is only comfortable in 3D. Not only that,
the wedge (outer product) is a product that creates volumes, hypervolumes, and can
also be applied to vectors, bivectors, trivectors, etc.

6.9   The Outer Product of Three 3D Vectors

Having seen that the outer product of two 3D vectors is represented by areal projections onto the three basis bivectors, let's explore the outer product of three 3D vectors.

Given

$$\mathbf{a} = a_1\mathbf{e}_1 + a_2\mathbf{e}_2 + a_3\mathbf{e}_3$$
$$\mathbf{b} = b_1\mathbf{e}_1 + b_2\mathbf{e}_2 + b_3\mathbf{e}_3$$
$$\mathbf{c} = c_1\mathbf{e}_1 + c_2\mathbf{e}_2 + c_3\mathbf{e}_3$$

then

$$\mathbf{a} \wedge \mathbf{b} \wedge \mathbf{c} = (a_1\mathbf{e}_1 + a_2\mathbf{e}_2 + a_3\mathbf{e}_3) \wedge (b_1\mathbf{e}_1 + b_2\mathbf{e}_2 + b_3\mathbf{e}_3) \wedge (c_1\mathbf{e}_1 + c_2\mathbf{e}_2 + c_3\mathbf{e}_3)$$
$$= [(a_1b_2 - a_2b_1)\mathbf{e}_1 \wedge \mathbf{e}_2 + (a_2b_3 - a_3b_2)\mathbf{e}_2 \wedge \mathbf{e}_3 + (a_3b_1 - a_1b_3)\mathbf{e}_3 \wedge \mathbf{e}_1]$$
$$\wedge (c_1\mathbf{e}_1 + c_2\mathbf{e}_2 + c_3\mathbf{e}_3).$$

At this stage we introduce another axiom: the outer product is associative. This means that $\mathbf{a} \wedge (\mathbf{b} \wedge \mathbf{c}) = (\mathbf{a} \wedge \mathbf{b}) \wedge \mathbf{c}$. Therefore, knowing that $\mathbf{a} \wedge \mathbf{a} = 0$:

$$\mathbf{a} \wedge \mathbf{b} \wedge \mathbf{c} = c_3(a_1b_2 - a_2b_1)\mathbf{e}_1 \wedge \mathbf{e}_2 \wedge \mathbf{e}_3 + c_1(a_2b_3 - a_3b_2)\mathbf{e}_2 \wedge \mathbf{e}_3 \wedge \mathbf{e}_1$$
$$+ c_2(a_3b_1 - a_1b_3)\mathbf{e}_3 \wedge \mathbf{e}_1 \wedge \mathbf{e}_2.$$

But we are left with the products $\mathbf{e}_1 \wedge \mathbf{e}_2 \wedge \mathbf{e}_3$, $\mathbf{e}_2 \wedge \mathbf{e}_3 \wedge \mathbf{e}_1$ and $\mathbf{e}_3 \wedge \mathbf{e}_1 \wedge \mathbf{e}_2$. Not to worry, because we know that $\mathbf{a} \wedge \mathbf{b} = -\mathbf{b} \wedge \mathbf{a}$. Therefore,

$$\mathbf{e}_2 \wedge \mathbf{e}_3 \wedge \mathbf{e}_1 = -\mathbf{e}_2 \wedge \mathbf{e}_1 \wedge \mathbf{e}_3 = \mathbf{e}_1 \wedge \mathbf{e}_2 \wedge \mathbf{e}_3$$

and

$$\mathbf{e}_3 \wedge \mathbf{e}_1 \wedge \mathbf{e}_2 = -\mathbf{e}_1 \wedge \mathbf{e}_3 \wedge \mathbf{e}_2 = \mathbf{e}_1 \wedge \mathbf{e}_2 \wedge \mathbf{e}_3.$$

Therefore, we can write $\mathbf{a} \wedge \mathbf{b} \wedge \mathbf{c}$ as

$$\mathbf{a} \wedge \mathbf{b} \wedge \mathbf{c} = c_3(a_1b_2 - a_2b_1)\mathbf{e}_1 \wedge \mathbf{e}_2 \wedge \mathbf{e}_3 + c_1(a_2b_3 - a_3b_2)\mathbf{e}_1 \wedge \mathbf{e}_2 \wedge \mathbf{e}_3$$
$$+ c_2(a_3b_1 - a_1b_3)\mathbf{e}_1 \wedge \mathbf{e}_2 \wedge \mathbf{e}_3$$

or

$$\mathbf{a} \wedge \mathbf{b} \wedge \mathbf{c} = [c_3(a_1b_2 - a_2b_1) + c_1(a_2b_3 - a_3b_2) + c_2(a_3b_1 - a_1b_3)]\mathbf{e}_1 \wedge \mathbf{e}_2 \wedge \mathbf{e}_3$$

or using a determinant:

$$\mathbf{a} \wedge \mathbf{b} \wedge \mathbf{c} = \begin{vmatrix} a_1 & b_1 & c_1 \\ a_2 & b_2 & c_2 \\ a_3 & b_3 & c_3 \end{vmatrix} \mathbf{e}_1 \wedge \mathbf{e}_2 \wedge \mathbf{e}_3$$

which is the well-known expression for the volume of a parallelepiped formed by three vectors.

The term $\mathbf{e}_1 \wedge \mathbf{e}_2 \wedge \mathbf{e}_3$ is a *trivector* and reminds us that the volume is oriented. If the sign of the determinant is positive, the original three vectors possess the same orientation of the three basis vectors. If the sign of the determinant is negative, the three vectors possess an orientation opposing that of the three basis vectors.

6.10   Axioms

One of the features of geometric algebra is that it behaves very similar to the everyday algebra of scalars:
Axiom 1: The associative rule:

$$\mathbf{a}(\mathbf{bc}) = (\mathbf{ab})\mathbf{c}.$$

Axiom 2: The left and right distributive rules:

$$\mathbf{a}(\mathbf{b} + \mathbf{c}) = \mathbf{ab} + \mathbf{ac}$$
$$(\mathbf{b} + \mathbf{c})\mathbf{a} = \mathbf{ba} + \mathbf{ca}.$$

The next four axioms describe how vectors interact with a scalar λ:
Axiom 3:

$$(\lambda \mathbf{a})\mathbf{b} = \lambda(\mathbf{ab}) = \lambda \mathbf{ab}.$$

Axiom 4:

$$\lambda(\phi \mathbf{a}) = (\lambda \phi)\mathbf{a}.$$

Axiom 5:

$$\lambda(\mathbf{a} + \mathbf{b}) = \lambda \mathbf{a} + \lambda \mathbf{b}.$$

Axiom 6:

$$(\lambda + \phi)\mathbf{a} = \lambda\mathbf{a} + \phi\mathbf{a}.$$

The next axiom that is adopted is
Axiom 7:

$$\mathbf{a}^2 = |\mathbf{a}|^2$$

which has already emerged as a consequence of the algebra. However, for non-Euclidean geometries, this can be set to $\mathbf{a}^2 = -|\mathbf{a}|^2$, which does not concern us here.

6.11 Notation

Having abandoned $\mathbf{i}, \mathbf{j}, \mathbf{k}$ for $\mathbf{e}_1, \mathbf{e}_2, \mathbf{e}_3$, it is convenient to convert geometric products $\mathbf{e}_1\mathbf{e}_2 \ldots \mathbf{e}_n$ to $\mathbf{e}_{12\ldots n}$. For example, $\mathbf{e}_1\mathbf{e}_2\mathbf{e}_3 = \mathbf{e}_{123}$. Furthermore, we must get used to the following substitutions:

$$\mathbf{e}_i\mathbf{e}_i\mathbf{e}_j = \mathbf{e}_j$$
$$\mathbf{e}_{21} = -\mathbf{e}_{12}$$
$$\mathbf{e}_{312} = \mathbf{e}_{123}$$
$$\mathbf{e}_{112} = \mathbf{e}_2$$
$$\mathbf{e}_{121} = -\mathbf{e}_2.$$

6.12 Grades, Pseudoscalars and Multivectors

As geometric algebra embraces such a wide range of objects, it is convenient to *grade* them as follows: scalars are grade 0, vectors are grade 1, bivectors are grade 2, and trivectors are grade 3, and so on for higher dimensions. In such a graded algebra it is traditional to call the highest grade element a *pseudoscalar*. Thus in 2D the pseudoscalar is $\mathbf{e}_{12}$ and in 3D the pseudoscalar is $\mathbf{e}_{123}$.

One very powerful feature of geometric algebra is the idea of a *multivector*, which is a linear combination of a scalar, vector, bivector, trivector or any other higher dimensional object. For example the following are multivectors:

$$\mathbf{A} = 3 + (2\mathbf{e}_1 + 3\mathbf{e}_2 + 4\mathbf{e}_3) + (5\mathbf{e}_{12} + 6\mathbf{e}_{23} + 7\mathbf{e}_{31}) + 8\mathbf{e}_{123}$$
$$\mathbf{B} = 2 + (2\mathbf{e}_1 + 2\mathbf{e}_2 + 3\mathbf{e}_3) + (4\mathbf{e}_{12} + 5\mathbf{e}_{23} + 6\mathbf{e}_{31}) + 7\mathbf{e}_{123}$$

and we can form their sum:

$$\mathbf{A} + \mathbf{B} = 5 + (4\mathbf{e}_1 + 5\mathbf{e}_2 + 7\mathbf{e}_3) + (9\mathbf{e}_{12} + 11\mathbf{e}_{23} + 13\mathbf{e}_{31}) + 15\mathbf{e}_{123}$$

or their difference:

$$\mathbf{A} - \mathbf{B} = 1 + (\mathbf{e}_2 + \mathbf{e}_3) + (\mathbf{e}_{12} + \mathbf{e}_{23} + \mathbf{e}_{31}) + \mathbf{e}_{123}.$$

We can even form their product $\mathbf{AB}$, but at the moment we have not explored the products between all these elements.

We can isolate any grade of a multivector using the following notation:

$$\langle multivector \rangle_g$$

where g identifies a particular grade. For example, say we have the following multivector:

$$2 + 3\mathbf{e}_1 + 2\mathbf{e}_2 - 5\mathbf{e}_{12} + 6\mathbf{e}_{123}$$

we extract the scalar term using:

$$\langle 2 + 3\mathbf{e}_1 + 2\mathbf{e}_2 - 5\mathbf{e}_{12} + 6\mathbf{e}_{123} \rangle_0 = 2$$

the vector term using

$$\langle 2 + 3\mathbf{e}_1 + 2\mathbf{e}_2 - 5\mathbf{e}_{12} + 6\mathbf{e}_{123} \rangle_1 = 3\mathbf{e}_1 + 2\mathbf{e}_2$$

the bivector term using:

$$\langle 2 + 3\mathbf{e}_1 + 2\mathbf{e}_2 - 5\mathbf{e}_{12} + 6\mathbf{e}_{123} \rangle_2 = -5\mathbf{e}_{12}$$

and the trivector term using:

$$\langle 2 + 3\mathbf{e}_1 + 2\mathbf{e}_2 - 5\mathbf{e}_{12} + 6\mathbf{e}_{123} \rangle_3 = 6\mathbf{e}_{123}.$$

It is also worth pointing out that the inner vector product converts two grade 1 elements, i.e. vectors, into a grade 0 element, i.e. a scalar, whereas the outer vector product converts two grade 1 elements into a grade 2 element, i.e. a bivector. Thus the inner product is a grade lowering operation, while the outer product is a grade raising operation. These qualities of the inner and outer products are associated with higher grade elements in the algebra. This is why the scalar product is renamed as the inner product, because the scalar product is synonymous with transforming vectors into scalars. Whereas, the inner product transforms two elements of grade n into a grade $n - 1$ element.

6.13   Redefining the Inner and Outer Products

As the geometric product is defined in terms of the inner and outer products, it seems only natural to expect that similar functions exist relating the inner and outer products in terms of the geometric product. Such functions do exist and emerge when we combine the following two equations:

$$\mathbf{ab} = \mathbf{a} \cdot \mathbf{b} + \mathbf{a} \wedge \mathbf{b} \tag{6.18}$$

$$\mathbf{ba} = \mathbf{a} \cdot \mathbf{b} - \mathbf{a} \wedge \mathbf{b}. \tag{6.19}$$

Adding and subtracting (6.18) and (6.19) we have

$$\mathbf{a} \cdot \mathbf{b} = \tfrac{1}{2}(\mathbf{ab} + \mathbf{ba}) \tag{6.20}$$

$$\mathbf{a} \wedge \mathbf{b} = \tfrac{1}{2}(\mathbf{ab} - \mathbf{ba}). \tag{6.21}$$

Equations (6.20) and (6.21) and used frequently to define the products between different grade elements.

6.14   The Inverse of a Vector

In traditional vector analysis we accept that it is impossible to divide by a vector, but that is not so in geometric algebra. In fact, we don't actually divide a multivector by another vector but find a way of representing the inverse of a vector. For example, we know that a unit vector $\hat{\mathbf{a}}$ is defined as

$$\hat{\mathbf{a}} = \frac{\mathbf{a}}{|\mathbf{a}|}$$

and using the geometric product

$$\hat{\mathbf{a}}^2 = \frac{\mathbf{a}^2}{|\mathbf{a}|^2} = 1$$

therefore,

$$\mathbf{b} = \frac{\mathbf{a}^2 \mathbf{b}}{|\mathbf{a}|^2}$$

and exploiting the associative nature of the geometric product we have

$$\mathbf{b} = \frac{\mathbf{a}(\mathbf{ab})}{|\mathbf{a}|^2}. \tag{6.22}$$

Equation (6.22) is effectively stating that, given the geometric product $\mathbf{ab}$ we can recover the vector $\mathbf{b}$ by pre-multiplying by $\mathbf{a}^{-1}$:

$$\mathbf{a}^{-1} = \frac{\mathbf{a}}{|\mathbf{a}|^2}.$$

Similarly, we can recover the vector $\mathbf{a}$ by post-multiplying by $\mathbf{b}^{-1}$:

$$\mathbf{a} = \frac{(\mathbf{ab})\mathbf{b}}{|\mathbf{b}|^2}.$$

For example, given two vectors

$$\mathbf{a} = \mathbf{e}_1 + 2\mathbf{e}_2$$
$$\mathbf{b} = 3\mathbf{e}_1 + 2\mathbf{e}_2$$

their geometric product is

$$\mathbf{ab} = 7 - 4\mathbf{e}_{12}.$$

Therefore, given $\mathbf{ab}$ and $\mathbf{a}$, we can recover $\mathbf{b}$ as follows:

$$\begin{aligned}
\mathbf{b} &= \frac{\mathbf{e}_1 + 2\mathbf{e}_2}{5}(7 - 4\mathbf{e}_{12}) \\
&= \tfrac{1}{5}(7\mathbf{e}_1 - 4\mathbf{e}_{112} + 14\mathbf{e}_2 - 8\mathbf{e}_{212}) \\
&= \tfrac{1}{5}(7\mathbf{e}_1 - 4\mathbf{e}_2 + 14\mathbf{e}_2 + 8\mathbf{e}_1) \\
\mathbf{b} &= 3\mathbf{e}_1 + 2\mathbf{e}_2.
\end{aligned}$$

Similarly, give $\mathbf{ab}$ and $\mathbf{b}$, $\mathbf{a}$ is recovered as follows:

$$\begin{aligned}
\mathbf{a} &= (7 - 4\mathbf{e}_{12})\frac{3\mathbf{e}_1 + 2\mathbf{e}_2}{13} \\
&= \tfrac{1}{13}(21\mathbf{e}_1 + 14\mathbf{e}_2 - 12\mathbf{e}_{121} - 8\mathbf{e}_{122}) \\
&= \tfrac{1}{13}(21\mathbf{e}_1 + 14\mathbf{e}_2 + 12\mathbf{e}_2 - 8\mathbf{e}_1) \\
\mathbf{a} &= \mathbf{e}_1 + 2\mathbf{e}_2.
\end{aligned}$$

Note that the inverse of a unit vector is the original vector:

$$\hat{\mathbf{a}}^{-1} = \frac{\hat{\mathbf{a}}}{|\hat{\mathbf{a}}|^2} = \hat{\mathbf{a}}.$$

6.15 The Imaginary Properties of the Outer Product

So far we know that the outer product of two vectors is represented by one or more unit basis vectors, such as

$$\mathbf{a} \wedge \mathbf{b} = \lambda_1 \mathbf{e}_{12} + \lambda_2 \mathbf{e}_{23} + \lambda_3 \mathbf{e}_{31}$$

where, in this case, the λ_i terms represent areas projected onto their respective unit basis bivectors. But what has not emerged is that the outer product is an imaginary quantity, which is revealed by expanding $\mathbf{e}_{12}^2$:

$$\mathbf{e}_{12}^2 = \mathbf{e}_{1212}$$

but as

$$\mathbf{e}_{21} = -\mathbf{e}_{12}$$

then

$$\mathbf{e}_{1(21)2} = -\mathbf{e}_{1(12)2}$$
$$= -\mathbf{e}_1^2 \mathbf{e}_2^2$$
$$\mathbf{e}_{12}^2 = -1.$$

Consequently, the geometric product effectively creates a complex number! Thus in a 2D scenario, given two vectors

$$\mathbf{a} = a_1 \mathbf{e}_1 + a_2 \mathbf{e}_2$$
$$\mathbf{b} = b_1 \mathbf{e}_1 + b_2 \mathbf{e}_2$$

their geometric product is

$$\mathbf{ab} = (a_1 b_1 + a_2 b_2) + (a_1 b_2 - a_2 b_1) \mathbf{e}_{12}$$

and knowing that $\mathbf{e}_{12} = i$, then we can write $\mathbf{ab}$ as

$$\mathbf{ab} = (a_1 b_1 + a_2 b_2) + (a_1 b_2 - a_2 b_1) i. \tag{6.23}$$

However, this notation is not generally adopted by the geometric community. The reason being that i is normally only associated with a scalar, with which it commutes. Whereas in 2D, $\mathbf{e}_{12}$ is associated with scalars and vectors, and although scalars present no problem, under some conditions, it anti-commutes with vectors. Consequently, an upper-case I is used so that there is no confusion between the two elements. Thus (6.23) is written as

Fig. 6.8 The effect of
pre-multiplying a vector by a
bivector

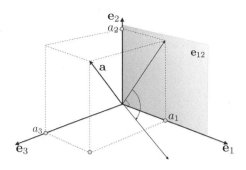

$$\mathbf{ab} = (a_1 b_1 + a_2 b_2) + (a_1 b_2 - a_2 b_1)I$$

where

$$I^2 = -1.$$

It goes without saying that the 3D unit basis bivectors are also imaginary quantities, so is $\mathbf{e}_{123}$.

Multiplying a complex number by i rotates it 90° on the complex plane. Therefore, it should be no surprise that multiplying a 2D vector by $\mathbf{e}_{12}$ rotates it by 90°. However, because vectors are sensitive to their product partners, we must remember that pre-multiplying a vector by $\mathbf{e}_{12}$ rotates a vector clockwise and post-multiplying rotates a vector anti-clockwise.

Whilst on the subject of rotations, let's consider what happens in 3D. We begin with a 3D vector

$$\mathbf{a} = a_1 \mathbf{e}_1 + a_2 \mathbf{e}_2 + a_3 \mathbf{e}_3$$

and the unit basis bivector $\mathbf{e}_{12}$ as shown in Fig. 6.8. Next we construct their geometric product by pre-multiplying $\mathbf{a}$ by $\mathbf{e}_{12}$:

$$\mathbf{e}_{12}\mathbf{a} = a_1 \mathbf{e}_{12}\mathbf{e}_1 + a_2 \mathbf{e}_{12}\mathbf{e}_2 + a_3 \mathbf{e}_{12}\mathbf{e}_3$$

which becomes

$$\mathbf{e}_{12}\mathbf{a} = a_1 \mathbf{e}_{121} + a_2 \mathbf{e}_{122} + a_3 \mathbf{e}_{123}$$
$$= -a_1 \mathbf{e}_2 + a_2 \mathbf{e}_1 + a_3 \mathbf{e}_{123}$$
$$= a_2 \mathbf{e}_1 - a_1 \mathbf{e}_2 + a_3 \mathbf{e}_{123}$$

and contains two parts: a vector $(a_2 \mathbf{e}_1 - a_1 \mathbf{e}_2)$ and a volume $a_3 \mathbf{e}_{123}$.

Figure 6.8 shows how the projection of vector $\mathbf{a}$ is rotated clockwise on the bivector $\mathbf{e}_{12}$. A volume is also created perpendicular to the bivector. This enables us to predict that if the vector is coplanar with the bivector, the entire vector is rotated $-90°$ and the volume component will be zero.

By post-multiplying $\mathbf{a}$ by $\mathbf{e}_{12}$ creates

$$\mathbf{a}\mathbf{e}_{12} = -a_2\mathbf{e}_1 + a_1\mathbf{e}_2 + a_3\mathbf{e}_{123}$$

which shows that while the volumetric element has remained the same, the projected vector is rotated anti-clockwise.

You may wish to show that the same happens with the other two bivectors.

6.16   Duality

The ability to exchange pairs of geometric elements such as lines and planes involves a *dual* operation, which in geometric algebra is relatively easy to define. For example, given a multivector $\mathbf{A}$ its dual $\mathbf{A}^*$ is defined as

$$\mathbf{A}^* = I\mathbf{A}$$

where I is the local pseudoscalar. For 2D this is $\mathbf{e}_{12}$ and for 3D it is $\mathbf{e}_{123}$. Therefore, given a 2D vector

$$\mathbf{a} = a_1\mathbf{e}_1 + a_2\mathbf{e}_2$$

its dual is

$$\begin{aligned}
\mathbf{a}^* &= \mathbf{e}_{12}(a_1\mathbf{e}_1 + a_2\mathbf{e}_2) \\
&= a_1\mathbf{e}_{121} + a_2\mathbf{e}_{122} \\
&= a_2\mathbf{e}_1 - a_1\mathbf{e}_2
\end{aligned}$$

which is another vector rotated 90° anti-clockwise.

It is easy to show that $(\mathbf{a}^*)^* = -\mathbf{a}$, and two further dual operations return the vector back to $\mathbf{a}$.

In 3D the dual of a vector $\mathbf{e}_1$ is

$$\mathbf{e}_{123}\mathbf{e}_1 = \mathbf{e}_{1231} = \mathbf{e}_{23}$$

which is the perpendicular bivector. Similarly, the dual of $\mathbf{e}_2$ is $\mathbf{e}_{31}$ and the dual of $\mathbf{e}_3$ is $\mathbf{e}_{12}$.

For a general vector $a_1\mathbf{e}_1 + a_2\mathbf{e}_2 + a_3\mathbf{e}_3$ its dual is

$$\mathbf{e}_{123}(a_1\mathbf{e}_1 + a_2\mathbf{e}_2 + a_3\mathbf{e}_3) = a_1\mathbf{e}_{1231} + a_2\mathbf{e}_{1232} + a_3\mathbf{e}_{1233}$$
$$= a_3\mathbf{e}_{12} + a_1\mathbf{e}_{23} + a_2\mathbf{e}_{31}.$$

The duals of the 3D basis bivectors are:

$$\mathbf{e}_{123}\mathbf{e}_{12} = \mathbf{e}_{12312} = -\mathbf{e}_3$$
$$\mathbf{e}_{123}\mathbf{e}_{23} = \mathbf{e}_{12323} = -\mathbf{e}_1$$
$$\mathbf{e}_{123}\mathbf{e}_{31} = \mathbf{e}_{12331} = -\mathbf{e}_2.$$

6.17 The Relationship between the Vector Product and the Outer Product

We have already discovered that there is a very close relationship between the vector product and the outer product, and just to recap: Given two vectors

$$\mathbf{a} = a_1\mathbf{e}_1 + a_2\mathbf{e}_2 + a_3\mathbf{e}_3$$
$$\mathbf{b} = b_1\mathbf{e}_1 + b_2\mathbf{e}_2 + b_3\mathbf{e}_3$$

then

$$\mathbf{a} \times \mathbf{b} = (a_2b_3 - a_3b_2)\mathbf{e}_1 + (a_3b_1 - a_1b_3)\mathbf{e}_2 + (a_1b_2 - a_2b_1)\mathbf{e}_3 \qquad (6.24)$$

and

$$\mathbf{a} \wedge \mathbf{b} = (a_2b_3 - a_3b_2)\mathbf{e}_2 \wedge \mathbf{e}_3 + (a_3b_1 - a_1b_3)\mathbf{e}_3 \wedge \mathbf{e}_1 + (a_1b_2 - a_2b_1)\mathbf{e}_1 \wedge \mathbf{e}_2$$

or

$$\mathbf{a} \wedge \mathbf{b} = (a_2b_3 - a_3b_2)\mathbf{e}_{23} + (a_3b_1 - a_1b_3)\mathbf{e}_{31} + (a_1b_2 - a_2b_1)\mathbf{e}_{12}. \qquad (6.25)$$

If we multiply (6.25) by I_{123} we obtain

$$I_{123}(\mathbf{a} \wedge \mathbf{b}) = (a_2b_3 - a_3b_2)\mathbf{e}_{123}\mathbf{e}_{23} + (a_3b_1 - a_1b_3)\mathbf{e}_{123}\mathbf{e}_{31} + (a_1b_2 - a_2b_1)\mathbf{e}_{123}\mathbf{e}_{12}$$
$$= -(a_2b_3 - a_3b_2)\mathbf{e}_1 - (a_3b_1 - a_1b_3)\mathbf{e}_2 - (a_1b_2 - a_2b_1)\mathbf{e}_3$$

which is identical to the cross product (6.24) apart from its sign. Therefore, we can state:

$$\mathbf{a} \times \mathbf{b} = -I_{123}(\mathbf{a} \wedge \mathbf{b}).$$

Table 6.1   Hamilton's quaternion product rules

	i	j	k
i	-1	k	$-j$
j	$-k$	-1	i
k	j	$-i$	-1

Table 6.2   3D bivector product rules

	$\mathbf{e}_{23}$	$\mathbf{e}_{31}$	$\mathbf{e}_{12}$
$\mathbf{e}_{23}$	-1	$-\mathbf{e}_{12}$	$\mathbf{e}_{31}$
$\mathbf{e}_{31}$	$\mathbf{e}_{12}$	-1	$-\mathbf{e}_{23}$
$\mathbf{e}_{12}$	$-\mathbf{e}_{31}$	$\mathbf{e}_{23}$	-1

Table 6.3   Left-handed 3D bivector product rules

	$\mathbf{e}_{32}$	$\mathbf{e}_{13}$	$\mathbf{e}_{21}$
$\mathbf{e}_{32}$	-1	$\mathbf{e}_{21}$	$-\mathbf{e}_{13}$
$\mathbf{e}_{13}$	$-\mathbf{e}_{21}$	-1	$\mathbf{e}_{32}$
$\mathbf{e}_{21}$	$\mathbf{e}_{13}$	$-\mathbf{e}_{32}$	-1

6.18   The Relationship between Quaternions and Bivectors

Hamilton's rules for the imaginaries i, j and k are shown in Table 6.1, whilst Table 6.2 shows the rules for 3D bivector products.

Although there is some agreement between the table entries, there is a sign reversal in some of them. However, if we switch to a left-handed axial system the bivectors become $\mathbf{e}_{32}, \mathbf{e}_{13}, \mathbf{e}_{21}$ and their products are as shown in Table 6.3.

If we now create a one-to-one correspondence (isomorphism) between the two systems:

$$i \leftrightarrow \mathbf{e}_{32} \quad j \leftrightarrow \mathbf{e}_{13} \quad k \leftrightarrow \mathbf{e}_{21}$$

there is a true correspondence between quaternions and a left-handed set of bivectors.

6.19   Reflections and Rotations

One of geometric algebra's strengths is the elegance it brings to calculating reflections and rotations. Unfortunately, there is insufficient space to examine the derivations of the formulae, but if you are interested, these can be found in the author's books [1, 2]. Let's start with 2D reflections.

Fig. 6.9 The reflection of a
2D vector

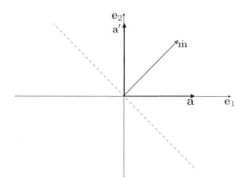

6.19.1   2D Reflections

Given a line, whose perpendicular unit vector is $\hat{\mathbf{m}}$ and a vector $\mathbf{a}$ its reflection $\mathbf{a}'$ is
given by

$$\mathbf{a}' = \hat{\mathbf{m}}\mathbf{a}\hat{\mathbf{m}}$$

which is rather elegant! For example, Fig. 6.9 shows a scenario where

$$\hat{\mathbf{m}} = \tfrac{1}{\sqrt{2}}(\mathbf{e}_1 + \mathbf{e}_2)$$
$$\mathbf{a} = \mathbf{e}_1$$

therefore,

$$\mathbf{a}' = \tfrac{1}{\sqrt{2}}(\mathbf{e}_1 + \mathbf{e}_2)(\mathbf{e}_1)\tfrac{1}{\sqrt{2}}(\mathbf{e}_1 + \mathbf{e}_2)$$
$$= \tfrac{1}{2}(1 - \mathbf{e}_{12})(\mathbf{e}_1 + \mathbf{e}_2)$$
$$= \tfrac{1}{2}(\mathbf{e}_1 + \mathbf{e}_2 + \mathbf{e}_2 - \mathbf{e}_1)$$
$$\mathbf{a}' = \mathbf{e}_2.$$

Note that in this scenario a reflection means a mirror image about the perpendicular
vector.

6.19.2   3D Reflections

Let's explore the 3D scenario shown in Fig. 6.10 where

Fig. 6.10 The reflection of a
3D vector

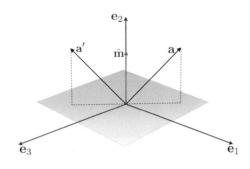

$$\mathbf{a} = \mathbf{e}_1 + \mathbf{e}_2 - \mathbf{e}_3$$
$$\hat{\mathbf{m}} = \mathbf{e}_2$$

therefore,

$$\mathbf{a}' = \mathbf{e}_2(\mathbf{e}_1 + \mathbf{e}_2 - \mathbf{e}_3)\mathbf{e}_2$$
$$= \mathbf{e}_{212} + \mathbf{e}_{222} - \mathbf{e}_{232}$$
$$= -\mathbf{e}_1 + \mathbf{e}_2 + \mathbf{e}_3.$$

As one might expect, it is also possible to reflect bivectors, trivectors and higher-dimensional objects, and for reasons of brevity, they are summarised as follows:

Reflecting about a line:

$$\begin{aligned}
\textit{scalars} \quad &\text{invariant}\\
\textit{vectors} \quad &\mathbf{v}' = \hat{\mathbf{m}}\mathbf{v}\hat{\mathbf{m}}\\
\textit{bivectors} \quad &\mathbf{B}' = \hat{\mathbf{m}}\mathbf{B}\hat{\mathbf{m}}\\
\textit{trivectors} \quad &\mathbf{T}' = \hat{\mathbf{m}}\mathbf{T}\hat{\mathbf{m}}.
\end{aligned}$$

Reflecting about a mirror:

$$\begin{aligned}
\textit{scalars} \quad &\text{invariant}\\
\textit{vectors} \quad &\mathbf{v}' = -\hat{\mathbf{m}}\mathbf{v}\hat{\mathbf{m}}\\
\textit{bivectors} \quad &\mathbf{B}' = \hat{\mathbf{m}}\mathbf{B}\hat{\mathbf{m}}\\
\textit{trivectors} \quad &\mathbf{T}' = -\hat{\mathbf{m}}\mathbf{T}\hat{\mathbf{m}}.
\end{aligned}$$

6.19.3 2D Rotations

Figure 6.11 shows a plan view of two mirrors M and N separated by an angle θ. The point P is in front of mirror M and subtends an angle α, and its reflection P_R exists in

Fig. 6.11 Rotating a point
by a double reflection

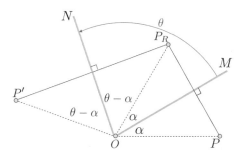

Fig. 6.12 Rotating a point
by a double reflection

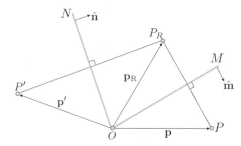

the virtual space behind M and also subtends an angle α with the mirror. The angle between P_R and N must be $\theta - \alpha$, and its reflection P' must also lie $\theta - \alpha$ behind N. By inspection, the angle between P and the double reflection P' is 2θ.

If we apply this double reflection transform to a collection of points, they are effectively all rotated 2θ about the origin where the mirrors intersect. The only slight drawback with this technique is that the angle of rotation is twice the angle between the mirrors.

Instead of using points, let's employ position vectors and substitute normal unit vectors for the mirrors' orientation. For example, Fig. 6.12 shows the same mirrors with unit normal vectors $\hat{\mathbf{m}}$ and $\hat{\mathbf{n}}$. After two successive reflections, P becomes P', and using the relationship:

$$\mathbf{v}' = -\hat{\mathbf{m}}\mathbf{v}\hat{\mathbf{m}}$$

we compute the reflections as follows:

$$\mathbf{p}_R = -\hat{\mathbf{m}}\mathbf{p}\hat{\mathbf{m}}$$
$$\mathbf{p}' = -\hat{\mathbf{n}}\mathbf{p}_R\hat{\mathbf{n}}$$
$$\mathbf{p}' = \hat{\mathbf{n}}\hat{\mathbf{m}}\mathbf{p}\hat{\mathbf{m}}\hat{\mathbf{n}}$$

which is also rather elegant and compact. However, we must remember that P is rotated twice the angle separating the mirrors, and the rotation is relative to the origin. Let's demonstrate this technique with an example.

Fig. 6.13 Rotating a point
by 180°

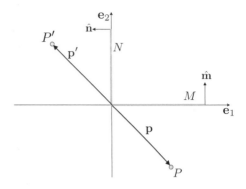

Figure 6.13 shows two mirrors M and N with unit normal vectors $\hat{\mathbf{m}}$, $\hat{\mathbf{n}}$ and
position vector $\mathbf{p}$:

$$\hat{\mathbf{m}} = \mathbf{e}_2$$
$$\hat{\mathbf{n}} = -\mathbf{e}_1$$
$$P = (1, -1)$$
$$\mathbf{p} = \mathbf{e}_1 - \mathbf{e}_2.$$

As the mirrors are separated by 90° the point P is rotated 180°:

$$\mathbf{p}' = \hat{\mathbf{n}}\hat{\mathbf{m}}\mathbf{p}\hat{\mathbf{m}}\hat{\mathbf{n}}$$
$$= -\mathbf{e}_1\mathbf{e}_2(\mathbf{e}_1 - \mathbf{e}_2)\mathbf{e}_2(-\mathbf{e}_1)$$
$$= \mathbf{e}_{12121} - \mathbf{e}_{12221}$$
$$= -\mathbf{e}_1 + \mathbf{e}_2$$
$$P' = (-1, 1).$$

6.20   Rotors

Quaternions are the natural choice for rotating vectors about an arbitrary axis, and
although it may not be immediately obvious, we have already started to discover
geometric algebra's equivalent.

We begin with

$$\mathbf{p}' = \hat{\mathbf{n}}\hat{\mathbf{m}}\mathbf{p}\hat{\mathbf{m}}\hat{\mathbf{n}}$$

and substitute $\mathbf{R}$ for $\hat{\mathbf{n}}\hat{\mathbf{m}}$ and $\tilde{\mathbf{R}}$ for $\hat{\mathbf{m}}\hat{\mathbf{n}}$, therefore,

$$\mathbf{p}' = \mathbf{R}\mathbf{p}\tilde{\mathbf{R}}$$

where $\mathbf{R}$ and $\tilde{\mathbf{R}}$ are called *rotors* which perform the same function as a quaternion.

Because geometric algebra is non-commutative, the sequence of elements, be they vectors, bivectors, trivectors, etc., is very important. Consequently, it is very useful to include a command that reverses a sequence of elements. The notation generally employed is the tilde " ˜ " symbol:

$$\mathbf{R} = \hat{\mathbf{n}}\hat{\mathbf{m}}$$
$$\tilde{\mathbf{R}} = \hat{\mathbf{m}}\hat{\mathbf{n}}.$$

Let's unpack a rotor in terms of its angle and bivector as follows:

The bivector defining the plane is $\hat{\mathbf{m}} \wedge \hat{\mathbf{n}}$ and θ is the angle between the vectors. Let

$$\mathbf{R} = \hat{\mathbf{n}}\hat{\mathbf{m}}$$
$$\tilde{\mathbf{R}} = \hat{\mathbf{m}}\hat{\mathbf{n}}$$

where

$$\hat{\mathbf{n}}\hat{\mathbf{m}} = \hat{\mathbf{n}} \cdot \hat{\mathbf{m}} - \hat{\mathbf{m}} \wedge \hat{\mathbf{n}}$$
$$\hat{\mathbf{m}}\hat{\mathbf{n}} = \hat{\mathbf{n}} \cdot \hat{\mathbf{m}} + \hat{\mathbf{m}} \wedge \hat{\mathbf{n}}$$
$$\hat{\mathbf{n}} \cdot \hat{\mathbf{m}} = \cos\theta$$
$$\hat{\mathbf{m}} \wedge \hat{\mathbf{n}} = \hat{\mathbf{B}}\sin\theta.$$

Therefore,

$$\mathbf{R} = \cos\theta - \hat{\mathbf{B}}\sin\theta$$
$$\tilde{\mathbf{R}} = \cos\theta + \hat{\mathbf{B}}\sin\theta.$$

We now have an equation that rotates a vector $\mathbf{p}$ through an angle 2θ about an axis defined by $\hat{\mathbf{B}}$:

$$\mathbf{p}' = \mathbf{R}\mathbf{p}\hat{\mathbf{R}}$$

or

$$\mathbf{p}' = \left[\cos\theta - \hat{\mathbf{B}}\sin\theta)\right]\mathbf{p}\left[\cos\theta + \hat{\mathbf{B}}\sin\theta\right].$$

We can also express this such that it identifies the real angle of rotation α:

$$\mathbf{p}' = \left[\cos(\alpha/2) - \hat{\mathbf{B}}\sin(\alpha/2)\right]\mathbf{p}\left[\cos(\alpha/2) + \hat{\mathbf{B}}\sin(\alpha/2)\right]. \qquad (6.26)$$

Fig. 6.14 Rotating a vector
by 90°

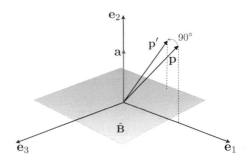

Equation (6.26) references a bivector, which may make you feel uncomfortable! But
remember, it simply identifies the axis perpendicular to its plane. Let's demonstrate
how (6.26) works with two examples.

Figure 6.14 shows a scenario where vector $\mathbf{p}$ is rotated 90° about $\mathbf{e}_2$ which is
perpendicular to $\hat{\mathbf{B}}$, where

$$\alpha = 90°$$
$$\mathbf{a} = \mathbf{e}_2$$
$$\mathbf{p} = \mathbf{e}_1 + \mathbf{e}_2$$
$$\hat{\mathbf{B}} = \mathbf{e}_{31}.$$

Therefore,

$$
\begin{aligned}
\mathbf{p}' &= (\cos 45° - \mathbf{e}_{31} \sin 45°)\,(\mathbf{e}_1 + \mathbf{e}_2)\,(\cos 45° + \mathbf{e}_{31} \sin 45°) \\
&= \left(\tfrac{\sqrt{2}}{2} - \tfrac{\sqrt{2}}{2}\mathbf{e}_{31}\right)(\mathbf{e}_1 + \mathbf{e}_2)\left(\tfrac{\sqrt{2}}{2} + \tfrac{\sqrt{2}}{2}\mathbf{e}_{31}\right) \\
&= \tfrac{1}{2}(\mathbf{e}_1 + \mathbf{e}_2 - \mathbf{e}_3 - \mathbf{e}_{312})(1 + \mathbf{e}_{31}) \\
&= \tfrac{1}{2}(\mathbf{e}_1 + \mathbf{e}_2 - \mathbf{e}_3 - \mathbf{e}_{312} - \mathbf{e}_3 - \mathbf{e}_{231} - \mathbf{e}_1 - \mathbf{e}_{31231}) \\
&= \tfrac{1}{2}(\mathbf{e}_1 + \mathbf{e}_2 - 2\mathbf{e}_3 - \mathbf{e}_1 + \mathbf{e}_2) \\
\mathbf{p}' &= \mathbf{e}_2 - \mathbf{e}_3.
\end{aligned}
$$

Observe what happens when the bivector's sign is reversed to $-\mathbf{e}_{31}$:

$$
\begin{aligned}
\mathbf{p}' &= (\cos 45° + \mathbf{e}_{31} \sin 45°)(\mathbf{e}_1 + \mathbf{e}_2)(\cos 45° - \mathbf{e}_{31} \sin 45°) \\
&= \left(\tfrac{\sqrt{2}}{2} + \tfrac{\sqrt{2}}{2}\mathbf{e}_{31}\right)(\mathbf{e}_1 + \mathbf{e}_2)\left(\tfrac{\sqrt{2}}{2} - \tfrac{\sqrt{2}}{2}\mathbf{e}_{31}\right) \\
&= \tfrac{1}{2}(\mathbf{e}_1 + \mathbf{e}_2 + \mathbf{e}_3 + \mathbf{e}_{312})(1 - \mathbf{e}_{31})
\end{aligned}
$$

Fig. 6.15 Rotating a vector by 120°

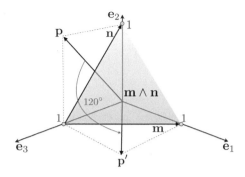

$$= \tfrac{1}{2}(\mathbf{e}_1 + \mathbf{e}_2 + \mathbf{e}_3 + \mathbf{e}_{312} + \mathbf{e}_3 - \mathbf{e}_{231} - \mathbf{e}_1 - \mathbf{e}_{31231})$$
$$= \tfrac{1}{2}(\mathbf{e}_1 + \mathbf{e}_2 + 2\mathbf{e}_3 - \mathbf{e}_1 + \mathbf{e}_2)$$
$$\mathbf{p}' = \mathbf{e}_2 + \mathbf{e}_3.$$

the rotation is clockwise about $\mathbf{e}_2$.

Figure 6.15 shows another scenario where vector $\mathbf{p}$ is rotated 120° about the bivector $\mathbf{B}$, where

$$\mathbf{m} = \mathbf{e}_1 - \mathbf{e}_3$$
$$\mathbf{n} = \mathbf{e}_2 - \mathbf{e}_3$$
$$\alpha = 120°$$
$$\mathbf{p} = \mathbf{e}_2 + \mathbf{e}_3$$
$$\mathbf{B} = \mathbf{m} \wedge \mathbf{n}$$
$$= (\mathbf{e}_1 - \mathbf{e}_3) \wedge (\mathbf{e}_2 - \mathbf{e}_3)$$
$$\mathbf{B} = \mathbf{e}_{12} + \mathbf{e}_{31} + \mathbf{e}_{23}.$$

Next, we normalise $\mathbf{B}$ to $\hat{\mathbf{B}}$:

$$\hat{\mathbf{B}} = \tfrac{1}{\sqrt{3}}(\mathbf{e}_{12} + \mathbf{e}_{23} + \mathbf{e}_{31})$$

therefore,

$$\mathbf{p}' = (\cos 60° - \hat{\mathbf{B}} \sin 60°)\mathbf{p}(\cos 60° + \hat{\mathbf{B}} \sin 60°)$$
$$= \left(\tfrac{1}{2} - \tfrac{1}{\sqrt{3}}(\mathbf{e}_{12} + \mathbf{e}_{23} + \mathbf{e}_{31})\tfrac{\sqrt{3}}{2}\right)(\mathbf{e}_2 + \mathbf{e}_3)\left(\tfrac{1}{2} + \tfrac{1}{\sqrt{3}}(\mathbf{e}_{12} + \mathbf{e}_{23} + \mathbf{e}_{31})\tfrac{\sqrt{3}}{2}\right)$$
$$= \left(\tfrac{1}{2} - \tfrac{\mathbf{e}_{12}}{2} - \tfrac{\mathbf{e}_{23}}{2} - \tfrac{\mathbf{e}_{31}}{2}\right)(\mathbf{e}_2 + \mathbf{e}_3)\left(\tfrac{1}{2} + \tfrac{\mathbf{e}_{12}}{2} + \tfrac{\mathbf{e}_{23}}{2} + \tfrac{\mathbf{e}_{31}}{2}\right)$$
$$= \tfrac{1}{4}(\mathbf{e}_2 + \mathbf{e}_3 - \mathbf{e}_1 - \mathbf{e}_{123} + \mathbf{e}_3 - \mathbf{e}_2 - \mathbf{e}_{312} + \mathbf{e}_1)(1 + \mathbf{e}_{12} + \mathbf{e}_{23} + \mathbf{e}_{31})$$
$$= \tfrac{1}{2}(\mathbf{e}_3 - \mathbf{e}_{123})(1 + \mathbf{e}_{12} + \mathbf{e}_{23} + \mathbf{e}_{31})$$

$$= \tfrac{1}{2}(\mathbf{e}_3 + \mathbf{e}_{312} - \mathbf{e}_2 + \mathbf{e}_1 - \mathbf{e}_{123} - \mathbf{e}_{12312} - \mathbf{e}_{12323} - \mathbf{e}_{12331})$$
$$= \tfrac{1}{2}(\mathbf{e}_3 - \mathbf{e}_2 + \mathbf{e}_1 + \mathbf{e}_3 + \mathbf{e}_1 + \mathbf{e}_2)$$
$$\mathbf{p}' = \mathbf{e}_1 + \mathbf{e}_3.$$

These examples show that rotors behave just like quaternions. Rotors not only rotate vectors, but they can be used to rotate bivectors, and even trivectors.

6.21 Summary

This chapter has provided a brief introduction to geometric algebra, which turns out to possess imaginary foundations. The algebra provides some elegant ways to compute reflections and rotations, which are extended in Chap. 10.

6.21.1 Summary of Formulae

2D Inner Product

$$\mathbf{a} = a_1\mathbf{e}_1 + a_2\mathbf{e}_2$$
$$\mathbf{b} = b_1\mathbf{e}_1 + b_2\mathbf{e}_2$$
$$\mathbf{a} \cdot \mathbf{b} = |\mathbf{a}||\mathbf{b}| \cos\theta = a_1b_1 + a_2b_2.$$

2D Outer Product

$$\mathbf{a} = a_1\mathbf{e}_1 + a_2\mathbf{e}_2$$
$$\mathbf{b} = b_1\mathbf{e}_1 + b_2\mathbf{e}_2$$
$$\mathbf{a} \wedge \mathbf{b} = |\mathbf{a}||\mathbf{b}| \sin\theta\mathbf{e}_1 \wedge \mathbf{e}_2 = (a_1b_2 - a_2b_1)\mathbf{e}_1 \wedge \mathbf{e}_2.$$

2D Geometric Product

$$\mathbf{a} = a_1\mathbf{e}_1 + a_2\mathbf{e}_2$$
$$\mathbf{b} = b_1\mathbf{e}_1 + b_2\mathbf{e}_2$$
$$\mathbf{ab} = |\mathbf{a}||\mathbf{b}| \cos\theta + |\mathbf{a}||\mathbf{b}| \sin\theta\mathbf{e}_1 \wedge \mathbf{e}_2 = (a_1b_1 + a_2b_2) + (a_1b_2 - a_2b_1)\mathbf{e}_1 \wedge \mathbf{e}_2.$$

3D Inner Product

$$\mathbf{a} = a_1\mathbf{e}_1 + a_2\mathbf{e}_2 + a_3\mathbf{e}_3$$
$$\mathbf{b} = b_1\mathbf{e}_1 + b_2\mathbf{e}_2 + b_3\mathbf{e}_3$$
$$\mathbf{a} \cdot \mathbf{b} = |\mathbf{a}||\mathbf{b}| \cos\theta = a_1b_1 + a_2b_2 + a_3b_3.$$

3D Outer Product

$$\mathbf{a} = a_1\mathbf{e}_1 + a_2\mathbf{e}_2 + a_3\mathbf{e}_3$$
$$\mathbf{b} = b_1\mathbf{e}_1 + b_2\mathbf{e}_2 + b_3\mathbf{e}_3$$
$$\mathbf{a} \wedge \mathbf{b} = (a_1b_2 - a_2b_1)\mathbf{e}_1 \wedge \mathbf{e}_2 + (a_2b_3 - a_3b_2)\mathbf{e}_2 \wedge \mathbf{e}_3 + (a_3b_1 - a_1b_3)\mathbf{e}_3 \wedge \mathbf{e}_1.$$

3D Geometric Product

$$\mathbf{a} = a_1\mathbf{e}_1 + a_2\mathbf{e}_2 + a_3\mathbf{e}_3$$
$$\mathbf{b} = b_1\mathbf{e}_1 + b_2\mathbf{e}_2 + b_3\mathbf{e}_3$$
$$\mathbf{ab} = \mathbf{a} \cdot \mathbf{b} + \mathbf{a} \wedge \mathbf{b} = (a_1b_1 + a_2b_2 + a_3b_3) +$$
$$(a_1b_2 - a_2b_1)\mathbf{e}_1 \wedge \mathbf{e}_2 + (a_2b_3 - a_3b_2)\mathbf{e}_2 \wedge \mathbf{e}_3 + (a_3b_1 - a_1b_3)\mathbf{e}_3 \wedge \mathbf{e}_1.$$

Outer Product of Three 3D Vectors

$$\mathbf{a} = a_1\mathbf{e}_1 + a_2\mathbf{e}_2 + a_3\mathbf{e}_3$$
$$\mathbf{b} = b_1\mathbf{e}_1 + b_2\mathbf{e}_2 + b_3\mathbf{e}_3$$
$$\mathbf{c} = c_1\mathbf{e}_1 + c_2\mathbf{e}_2 + c_3\mathbf{e}_3$$
$$\mathbf{a} \wedge \mathbf{b} \wedge \mathbf{c} = [c_3(a_1b_2 - a_2b_1) + c_1(a_2b_3 - a_3b_2) + c_2(a_3b_1 - a_1b_3)]\,\mathbf{e}_1 \wedge \mathbf{e}_2 \wedge \mathbf{e}_3$$
$$\mathbf{a} \wedge \mathbf{b} \wedge \mathbf{c} = \begin{vmatrix} a_1 & b_1 & c_1 \\ a_2 & b_2 & c_2 \\ a_3 & b_3 & c_3 \end{vmatrix} \mathbf{e}_1 \wedge \mathbf{e}_2 \wedge \mathbf{e}_3.$$

Generalised Inner and Outer Products

$$\mathbf{ab} = \mathbf{a} \cdot \mathbf{b} + \mathbf{a} \wedge \mathbf{b}$$
$$\mathbf{ba} = \mathbf{a} \cdot \mathbf{b} - \mathbf{a} \wedge \mathbf{b}$$
$$\mathbf{a} \cdot \mathbf{b} = \tfrac{1}{2}(\mathbf{ab} + \mathbf{ba})$$
$$\mathbf{a} \wedge \mathbf{b} = \tfrac{1}{2}(\mathbf{ab} - \mathbf{ba}).$$

Inverse of a Vector

$$\mathbf{a}^{-1} = \frac{\mathbf{a}}{|\mathbf{a}|^2}.$$

Imaginary Properties of the Outer Product

$$\mathbf{a} \wedge \mathbf{b} = \lambda_1\mathbf{e}_{12} + \lambda_2\mathbf{e}_{23} + \lambda_3\mathbf{e}_{31}$$
$$\mathbf{e}_{12}^2 = \mathbf{e}_{23}^2 = \mathbf{e}_{31}^2 = -1.$$

Reflections
About a line:

$$\begin{aligned} vectors \quad \mathbf{v}' &= \hat{\mathbf{m}}\mathbf{v}\hat{\mathbf{m}} \\ bivectors \quad \mathbf{B}' &= \hat{\mathbf{m}}\mathbf{B}\hat{\mathbf{m}} \\ trivectors \quad \mathbf{T}' &= \hat{\mathbf{m}}\mathbf{T}\hat{\mathbf{m}}. \end{aligned}$$

In a mirror:

$$\begin{aligned} vectors \quad \mathbf{v}' &= -\hat{\mathbf{m}}\mathbf{v}\hat{\mathbf{m}} \\ bivectors \quad \mathbf{B}' &= \hat{\mathbf{m}}\mathbf{B}\hat{\mathbf{m}} \\ trivectors \quad \mathbf{T}' &= -\hat{\mathbf{m}}\mathbf{T}\hat{\mathbf{m}}. \end{aligned}$$

6.22   Worked Examples

6.22.1   2D Inner Product

Calculate the inner product of $\mathbf{a} = 1\mathbf{e}_1$ and $\mathbf{b} = 1\mathbf{e}_1 + 2\mathbf{e}_2$.
Solution: The inner product is given by

$$\begin{aligned} \mathbf{a} \cdot \mathbf{b} &= a_1 b_1 + a_2 b_2 \\ &= 1 \times 1 + 0 \times 2 = 1. \end{aligned}$$

6.22.2   2D Outer Product

Calculate the outer product of $\mathbf{a} = 1\mathbf{e}_1$ and $\mathbf{b} = 1\mathbf{e}_1 + 2\mathbf{e}_2$.

Solution: The outer product is given by

$$\begin{aligned} \mathbf{a} \wedge \mathbf{b} &= (a_1 b_2 - a_2 b_1)\mathbf{e}_1 \wedge \mathbf{e}_2 \\ &= (1 \times 2 - 0 \times 1)\mathbf{e}_1 \wedge \mathbf{e}_2 = 2\mathbf{e}_1 \wedge \mathbf{e}_2. \end{aligned}$$

6.22.3   2D Geometric Product

Calculate the geometric product of $\mathbf{a} = \mathbf{e}_1$ and $\mathbf{b} = \mathbf{e}_1 + \mathbf{e}_2$.
Solution: The geometric product is given by

$$\mathbf{ab} = \mathbf{a} \cdot \mathbf{b} + \mathbf{a} \wedge \mathbf{b}$$
$$\mathbf{a} \cdot \mathbf{b} = 1$$
$$\mathbf{a} \wedge \mathbf{b} = 2\mathbf{e}_1 \wedge \mathbf{e}_2$$
$$\mathbf{ab} = 1 + 2\mathbf{e}_1 \wedge \mathbf{e}_2.$$

6.22.4 3D Inner Product

Calculate the inner product of $\mathbf{a} = \mathbf{e}_1 + 2\mathbf{e}_2 - 3\mathbf{e}_3$ and $\mathbf{b} = 2\mathbf{e}_1 - 3\mathbf{e}_2 + 4\mathbf{e}_3$.
Solution: The inner product is given by

$$\mathbf{a} \cdot \mathbf{b} = a_1b_1 + a_2b_2 + a_3b_3$$
$$= 1 \times 2 + 2 \times (-3) - 3 \times 4$$
$$= -16.$$

6.22.5 3D Outer Product

Calculate the outer product of $\mathbf{a} = \mathbf{e}_1 + 2\mathbf{e}_2 + 3\mathbf{e}_3$ and $\mathbf{b} = 2\mathbf{e}_1 + 3\mathbf{e}_2 + 5\mathbf{e}_3$.
Solution: The outer product is given by

$$\mathbf{a} \wedge \mathbf{b} = (a_1b_2 - a_2b_1)\mathbf{e}_1 \wedge \mathbf{e}_2 + (a_2b_3 - a_3b_2)\mathbf{e}_2 \wedge \mathbf{e}_3 + (a_3b_1 - a_1b_3)\mathbf{e}_3 \wedge \mathbf{e}_1$$
$$= (1 \times 3 - 2 \times 2)\mathbf{e}_1 \wedge \mathbf{e}_2 + (2 \times 5 - 3 \times 3)\mathbf{e}_2 \wedge \mathbf{e}_3 + (3 \times 2 - 1 \times 5)\mathbf{e}_2 \wedge \mathbf{e}_3$$
$$= -1\mathbf{e}_1 \wedge \mathbf{e}_2 + 1\mathbf{e}_2 \wedge \mathbf{e}_3 + 1\mathbf{e}_2 \wedge \mathbf{e}_3.$$

6.22.6 3D Geometric Product

Calculate the geometric product of $\mathbf{a} = \mathbf{e}_1 + 2\mathbf{e}_2 + 3\mathbf{e}_3$ and $\mathbf{b} = 2\mathbf{e}_1 + 3\mathbf{e}_2 + 5\mathbf{e}_3$.

Solution: The geometric product is given by

$$\mathbf{ab} = \mathbf{a} \cdot \mathbf{b} + \mathbf{a} \wedge \mathbf{b}$$
$$= -16 - 1\mathbf{e}_1 \wedge \mathbf{e}_2 + 1\mathbf{e}_2 \wedge \mathbf{e}_3 + 1\mathbf{e}_2 \wedge \mathbf{e}_3.$$

6.22.7   Outer Product of Three Vectors

Calculate the outer product of

$$\mathbf{a} = 1\mathbf{e}_1 + 1\mathbf{e}_2 + 1\mathbf{e}_3$$
$$\mathbf{b} = 2\mathbf{e}_1 + 4\mathbf{e}_2 + 3\mathbf{e}_3$$
$$\mathbf{c} = -5\mathbf{e}_1 + 6\mathbf{e}_2 + 7\mathbf{e}_3.$$

Solution: The outer product of three vectors is given by

$$\mathbf{a} \wedge \mathbf{b} \wedge \mathbf{c} = \begin{vmatrix} a_1 & b_1 & c_1 \\ a_2 & b_2 & c_2 \\ a_3 & b_3 & c_3 \end{vmatrix} \mathbf{e}_1 \wedge \mathbf{e}_2 \wedge \mathbf{e}_3$$

$$= \begin{vmatrix} 1 & 1 & 1 \\ 2 & 4 & 3 \\ -5 & 6 & 7 \end{vmatrix} \mathbf{e}_1 \wedge \mathbf{e}_2 \wedge \mathbf{e}_3$$

$$= [(4 \times 7) + (3 \times -5) + (2 \times 6) - (3 \times 6) - (2 \times 7) - (4 \times -5)]\mathbf{e}_1 \wedge \mathbf{e}_2 \wedge \mathbf{e}_3$$

$$= (28 - 15 + 12 - 18 - 14 + 20)\mathbf{e}_1 \wedge \mathbf{e}_2 \wedge \mathbf{e}_3$$

$$= 13\mathbf{e}_1 \wedge \mathbf{e}_2 \wedge \mathbf{e}_3.$$

6.22.8   Inverse of a Vector

Find the inverse of $\mathbf{a} = 1\mathbf{e}_1 + 2\mathbf{e}_2 + 3\mathbf{e}_3$.

Solution: The inverse of a vector is given by

$$\mathbf{a}^{-1} = \mathbf{a}/|\mathbf{a}|^2$$
$$|\mathbf{a}| = \sqrt{1^2 + 2^2 + 3^2}$$
$$|\mathbf{a}|^2 = 14$$
$$\mathbf{a}^{-1} = \tfrac{1}{14}(1\mathbf{e}_1 + 2\mathbf{e}_2 + 3\mathbf{e}_3).$$

6.22.9   Recovering a Vector from a Geometric Product

Find $\mathbf{b}$, given $\mathbf{ab} = 3 + \mathbf{e}_{12}$ and $\mathbf{a} = 2\mathbf{e}_1 + \mathbf{e}_2$.

Solution: Premultiply $\mathbf{ab}$ by $\mathbf{a}^{-1}$.

$$|\mathbf{a}|^2 = 5$$
$$\mathbf{a}^{-1} = \tfrac{1}{5}(2\mathbf{e}_1 + \mathbf{e}_2)$$
$$\mathbf{b} = \tfrac{1}{5}(2\mathbf{e}_1 + \mathbf{e}_2)(3 + \mathbf{e}_{12})$$
$$= \tfrac{1}{5}(6\mathbf{e}_1 + 2\mathbf{e}_{112} + 3\mathbf{e}_2 + \mathbf{e}_{212})$$
$$= \tfrac{1}{5}(6\mathbf{e}_1 + 2\mathbf{e}_2 + 3\mathbf{e}_2 - \mathbf{e}_1)$$
$$= \mathbf{e}_1 + \mathbf{e}_2.$$

6.22.10   Reflecting a 2D Vector about a Line

Calculate the reflection of $\mathbf{v} = \mathbf{e}_1 + \mathbf{e}_2$ about the normal $\hat{\mathbf{m}} = \mathbf{e}_2$.

Solution: The reflected vector is given by

$$\mathbf{v}' = \hat{\mathbf{m}}\mathbf{v}\hat{\mathbf{m}}$$
$$= \mathbf{e}_2(\mathbf{e}_1 + \mathbf{e}_2)\mathbf{e}_2$$
$$= (\mathbf{e}_{21} + 1)\mathbf{e}_2$$
$$= \mathbf{e}_{212} + \mathbf{e}_2$$
$$= -\mathbf{e}_1 + \mathbf{e}_2.$$

6.22.11   Reflecting a 3D Vector about a Line

Calculate the reflection of $\mathbf{v} = \mathbf{e}_1 + \mathbf{e}_2 + \mathbf{e}_3$ about the normal $\hat{\mathbf{m}} = \mathbf{e}_2$.

Solution: The reflected vector is given by

$$\mathbf{v}' = \hat{\mathbf{m}}\mathbf{v}\hat{\mathbf{m}}$$
$$= \mathbf{e}_2(\mathbf{e}_1 + \mathbf{e}_2 + \mathbf{e}_3)\mathbf{e}_2$$
$$= (\mathbf{e}_{21} + 1 + \mathbf{e}_{23})\mathbf{e}_2$$
$$= \mathbf{e}_{212} + \mathbf{e}_2 + \mathbf{e}_{232}$$
$$= -\mathbf{e}_1 + \mathbf{e}_2 - \mathbf{e}_3.$$

6.22.12   *Rotating a 3D Vector*

Rotate vector $\mathbf{p} = 2\mathbf{e}_1 + \mathbf{e}_2$ by $\alpha = 90°$ about $\mathbf{e}_1$ which is perpendicular to $\hat{\mathbf{B}} = \mathbf{e}_{23}$.

Solution: The rotated vector is given by

$$\mathbf{p}' = \left[\cos(\alpha/2) - \hat{\mathbf{B}}\sin(\alpha/2)\right]\mathbf{p}\left[\cos(\alpha/2) + \hat{\mathbf{B}}\sin(\alpha/2)\right]$$
$$= \left[\cos 45° - \mathbf{e}_{23}\sin 45°\right](2\mathbf{e}_1 + \mathbf{e}_2)\left[\cos 45° + \mathbf{e}_{23}\sin 45°\right]$$
$$= \tfrac{1}{2}\left[2\mathbf{e}_1 + \mathbf{e}_2 - 2\mathbf{e}_{231} - \mathbf{e}_{232}\right][1 + \mathbf{e}_{23}]$$
$$= \tfrac{1}{2}\left[2\mathbf{e}_1 + \mathbf{e}_2 + \mathbf{e}_3 - 2\mathbf{e}_{123}\right][1 + \mathbf{e}_{23}]$$
$$= \tfrac{1}{2}\left[2\mathbf{e}_1 + \mathbf{e}_2 + \mathbf{e}_3 - 2\mathbf{e}_{123} + 2\mathbf{e}_{123} + \mathbf{e}_{223} + \mathbf{e}_{323} - 2\mathbf{e}_{12323}\right]$$
$$= \tfrac{1}{2}\left[2\mathbf{e}_1 + 2\mathbf{e}_3 + 2\mathbf{e}_1\right]$$
$$= \left[2\mathbf{e}_1 + \mathbf{e}_3\right].$$

References

1. Vince JA (2008) Geometric algebra for computer graphics. Springer, Berlin
2. Vince JA (2009) Geometric algebra: an algebraic system for computer games and animation. Springer, Berlin
3. Grassmann HG (1840) Theorie der Ebbe und Flut, prfungsarbeit 1840 und Abhandlungen zur mathematischen. Physik aus dem Nachlasse, vol 3, Part 1

Chapter 7
Trigonometric Identities Using Complex Numbers

7.1 Introduction

Proving trigonometric identities is normally approached geometrically, where one constructs a diagram containing useful ratios, to reveal the required answer using logic. In this chapter I show how complex numbers provide an algebraic way of proving trigonometric identities.

7.2 Compound Angle Identities

Compound angle identities such as $\sin(\alpha + \beta)$ are normally taught using Euclidean geometry. For example, Fig. 7.1 does not seem to be a natural answer to the problem, nevertheless, is used to tackle the problem as follows.

$$\sin(\alpha + \beta) = \frac{FD}{AD} = \frac{BC + ED}{AD}$$
$$= \frac{BC}{AD}\frac{AC}{AC} + \frac{ED}{AD}\frac{CD}{CD}$$
$$= \frac{BC}{AC}\frac{AC}{AD} + \frac{ED}{CD}\frac{CD}{AD}$$
$$\sin(\alpha + \beta) = \sin\alpha\cos\beta + \cos\alpha\sin\beta. \tag{7.1}$$

Now let's expand $\cos(\alpha + \beta)$ with reference to Fig. 7.1:

$$\cos(\alpha + \beta) = \frac{AE}{AD} = \frac{AB - EC}{AD}$$
$$= \frac{AB}{AD}\frac{AC}{AC} - \frac{EC}{AD}\frac{CD}{CD}$$

© Springer International Publishing AG, part of Springer Nature 2018
J. Vince, *Imaginary Mathematics for Computer Science*,
https://doi.org/10.1007/978-3-319-94637-5_7

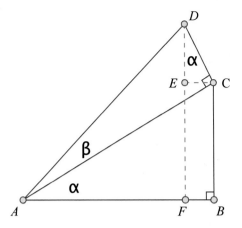

Fig. 7.1 The geometry to expand $\sin(\alpha + \beta)$

$$= \frac{AB}{AC}\frac{AC}{AD} - \frac{EC}{CD}\frac{CD}{AD}$$

$$\cos(\alpha + \beta) = \cos\alpha\cos\beta - \sin\alpha\sin\beta. \tag{7.2}$$

However, we can derive (7.1) and (7.2) by employing complex numbers as follows.

$$\begin{aligned}
\cos(\theta + \beta) + i\sin(\theta + \beta) &= e^{i(\theta+\beta)} = e^{i\theta}e^{i\beta}\\
&= (\cos\theta + i\sin\theta)(\cos\beta + i\sin\beta)\\
&= (\cos\theta\cos\beta - \sin\theta\sin\beta) + i(\sin\theta\cos\beta + \cos\theta\sin\beta)
\end{aligned}$$

equating real and imaginary parts, we have

$$\cos(\theta + \beta) = \cos\theta\cos\beta - \sin\theta\sin\beta$$
$$\sin(\theta + \beta) = \sin\theta\cos\beta + \cos\theta\sin\beta$$

and if we make β negative, we obtain

$$\cos(\theta - \beta) = \cos\theta\cos\beta + \sin\theta\sin\beta$$
$$\sin(\theta - \beta) = \sin\theta\cos\beta - \cos\theta\sin\beta.$$

This is much more intuitive, and just relies on the relationship

$$e^{i\alpha} = \cos\alpha + i\sin\alpha$$

furthermore, no diagram is required!

7.3   de Moivre's Theorem

de Moivre's theorem states

$$(\cos\theta + i\sin\theta)^n = \cos(n\theta) + i\sin(n\theta).$$

which we have seen is useful to deduce various trigonometric identities. However, we can isolate $\cos^n\theta$ and $\sin^n\theta$ as follows:

$$\cos^n\theta = \left(\frac{e^{i\theta} + e^{-i\theta}}{2}\right)^n$$

$$\sin^n\theta = \left(\frac{e^{i\theta} - e^{-i\theta}}{2i}\right)^n$$

which, using the binomial expansion, can be used to deduce identities of the form $\cos^n\theta$ and $\sin^n\theta$. Let's evaluate $\cos^n\theta$ and $\sin^n\theta$, $2 \le n \le 5$.

$$\cos^2\theta = \left(\frac{e^{i\theta} + e^{-i\theta}}{2}\right)^2$$

$$= \frac{e^{i2\theta} + 2 + e^{-i2\theta}}{4}$$

$$= \tfrac{1}{2}\left(\frac{2 + e^{i2\theta} + e^{-i2\theta}}{2}\right)$$

$$\cos^2\theta = \frac{1 + \cos(2\theta)}{2}.$$

$$\sin^2\theta = \left(\frac{e^{i\theta} - e^{-i\theta}}{2i}\right)^2$$

$$= \frac{e^{i2\theta} - 2 + e^{-i2\theta}}{-4}$$

$$= \tfrac{1}{2}\left(\frac{2 - (e^{i2\theta} + e^{-i2\theta})}{2}\right)$$

$$\sin^2\theta = \frac{1 - \cos(2\theta)}{2}.$$

$$\cos^3\theta = \left(\frac{e^{i\theta} + e^{-i\theta}}{2}\right)^3$$

$$= \frac{e^{i3\theta} + 3e^{i\theta} + 3e^{-i\theta} + e^{-i3\theta}}{8}$$

$$= \tfrac{1}{4}\left(\frac{3(e^{i\theta} + e^{-i\theta}) + e^{i3\theta} + e^{-3\theta}}{2}\right)$$

$$\cos^3\theta = \frac{3\cos\theta + \cos(3\theta)}{4}.$$

$$\sin^3\theta = \left(\frac{e^{i\theta} - e^{-i\theta}}{2i}\right)^3$$

$$= \frac{e^{i3\theta} - 3e^{i\theta} + 3e^{-i\theta} - e^{-i3\theta}}{-8i}$$

$$= \frac{1}{4}\left(\frac{(e^{i3\theta} - e^{-3\theta}) - 3(e^{i\theta} - e^{-i\theta})}{-2i}\right)$$

$$\sin^3\theta = \frac{3\sin\theta - \sin(3\theta)}{4}.$$

$$\cos^4\theta = \left(\frac{e^{i\theta} + e^{-i\theta}}{2}\right)^4$$

$$= \frac{e^{i4\theta} + 4e^{i2\theta} + 6 + 4e^{-i2\theta} + e^{-i4\theta}}{16}$$

$$= \frac{1}{8}\left(\frac{6 + 4(e^{i2\theta} + e^{-i2\theta}) + (e^{i4\theta} + e^{-i4\theta})}{2}\right)$$

$$\cos^4\theta = \frac{3 + 4\cos(2\theta) + \cos(4\theta)}{8}.$$

$$\sin^4\theta = \left(\frac{e^{i\theta} - e^{-i\theta}}{2i}\right)^4$$

$$= \frac{e^{i4\theta} - 4e^{i2\theta} + 6 - 4e^{-i2\theta} + e^{-i4\theta}}{16}$$

$$= \frac{1}{8}\left(\frac{6 - 4(e^{i2\theta} + e^{-i2\theta}) + (e^{i4\theta} + e^{-i4\theta})}{2}\right)$$

$$\sin^4\theta = \frac{3 - 4\cos(2\theta) + \cos(4\theta)}{8}.$$

$$\cos^5\theta = \left(\frac{e^{i\theta} + e^{-i\theta}}{2}\right)^5$$

$$= \frac{e^{i5\theta} + 5e^{i3\theta} + 10e^{i\theta} + 10e^{-i\theta} + 5e^{-i3\theta} + e^{i5\theta}}{32}$$

$$= \frac{1}{16}\left(\frac{10(e^{i\theta} + e^{-i\theta}) + 5(e^{i3\theta} + e^{-i3\theta}) + (e^{i5\theta} + e^{-i5\theta})}{2}\right)$$

$$\cos^5\theta = \frac{10\cos\theta + 5\cos(3\theta) + \cos(5\theta)}{16}.$$

$$\sin^5\theta = \left(\frac{e^{i\theta} - e^{-i\theta}}{2i}\right)^5$$

$$= \frac{e^{i5\theta} - 5e^{i3\theta} + 10e^{i\theta} - 10e^{-i\theta} + 5e^{-i3\theta} - e^{i5\theta}}{32i}$$

$$= \frac{1}{16}\left(\frac{10(e^{i\theta}-e^{-i\theta})-5(e^{i3\theta}-e^{-i3\theta})+(e^{i5\theta}-e^{-i5\theta})}{2i}\right)$$

$$\sin^5\theta = \frac{10\sin\theta - 5\sin(3\theta) + \sin(5\theta)}{16}.$$

7.4 Summary

Although $e^{i\theta}$ does not appear to be a natural mathematical object for proving trigonometric identities, it does provide some elegant proofs for compound angles and identities of the form $\cos^n\theta$ and $\sin^n\theta$. Unfortunately, we often discover complex numbers just when we have mastered geometric solutions to these problems!

Chapter 8
Combining Waves Using Complex Numbers

8.1 Introduction

Waves play an important role in transferring energy and information from one place to another, whether they are waves of air pressure, water, magnetism, electricity or gravity. When waves are combined, they cancel and reinforce one another; and as waves possess velocity, wavelength, frequency, amplitude and phase, computing the resultant waveform can be difficult using conventional trigonometric functions. Fortunately, complex numbers greatly simplify the analysis, and this chapter explains how.

8.2 Wave Equation

Before we begin, let's agree the notation used to describe sinusoidal waves. The following terms are often used to describe a sinusoidal waveform:

$$k = \text{the angular wavenumber [rad/metre]}$$
$$\omega = \text{the angular frequency [rad/sec]}$$
$$\lambda = 2\pi/k = \text{the wavelength [metre]}$$
$$\upsilon = \omega/2\pi = \text{the frequency [cycles/sec].}$$

A sinusoidal wave is written

$$\psi(x) = a\cos(kx)$$

where ψ is the Greek letter psi, a is the amplitude, k is the angular wavenumber, and x is a position along the wave.

© Springer International Publishing AG, part of Springer Nature 2018
J. Vince, *Imaginary Mathematics for Computer Science*,
https://doi.org/10.1007/978-3-319-94637-5_8

Fig. 8.1 A cosine wave

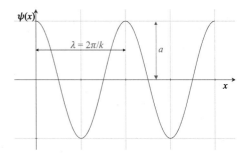

Figure 8.1 shows a cosine wave with amplitude a, wavelength $\lambda = 2\pi/k$, and the distance x in metres. If $k = 1$ [rad/metre], then $\lambda = 2\pi$ [metre].

If the wave changes with time, the parameter t must be multiplied by ω to create units of radians:

$$\psi(x, t) = a\cos(kx - \omega t).$$

To determine the phase speed c_p, we increment $t = t + 1$, and $x = x + \omega/k$:

$$\psi(x, t) = a\cos\left(k\left(x + \frac{\omega}{k}\right) - \omega(t + 1)\right)$$
$$= a\cos(kx + \omega - \omega t - \omega)$$
$$= a\cos(kx - \omega t).$$

Therefore, the wave's phase speed is the ratio of the increments:

$$c_p = \frac{\omega/k}{1} = \frac{\omega}{k}.$$

Next, we need to introduce a phase angle term ϕ, in order to delay or advance the wave:

$$\psi(x, t, \phi) = a\cos(kx - \omega t + \phi). \tag{8.1}$$

8.3   Combining Waves

We can now use trigonometric identities or complex numbers to resolve various waveform combinations with different frequency, phase and amplitude. However, this entails many combinations:

$$a \sin x + b \sin x$$
$$a \sin x + b \cos x$$
$$a \sin x + b \cos(x + \phi)$$
$$a \cos x + b \sin(x + \phi)$$
$$a \sin(x + \phi) + b \cos(x + \alpha)$$

etc.

So rather than develop a solution for every combination, I will illustrate a strategy, that will permit the reader to undertake other waveform combinations. Let's begin with trigonometric identities.

8.3.1 Using Trigonometric Identities

In the following description, either the cosine or sine function could be used, but I have chosen the cosine function. Consequently, the following identities will be useful

$$\cos(\alpha + \beta) = \cos \alpha \cos \beta - \sin \alpha \sin \beta$$
$$\cos \alpha + \cos \beta = 2 \cos \left(\frac{\alpha + \beta}{2} \right) \cos \left(\frac{\alpha - \beta}{2} \right).$$

To simplify the following waveforms, β is substituted for $kx - \omega t$. The simplest combination is two waves with the same frequency and amplitude, but no phase difference:

$$\psi_1 = a \cos \beta$$
$$\psi_2 = a \cos \beta$$
$$\psi_1 + \psi_2 = 2a \cos \beta.$$

Next, two waves with the same frequency and no phase difference, but different amplitudes:

$$\psi_1 = a_1 \cos \beta$$
$$\psi_2 = a_2 \cos \beta$$
$$\psi_1 + \psi_2 = (a_1 + a_2) \cos \beta.$$

Next, two waves with the same frequency and amplitude, but one includes a phase change:

$$\psi_1 = a \cos(\beta + \phi)$$
$$\psi_2 = a \cos \beta.$$

The aim is to create a wave equation of the form $A \cos \Phi$, where A is the amplitude, and references the original amplitude a and phase angle ϕ, and Φ references the original frequency and phase angle:

$$\psi_1 + \psi_2 = a \cos(\beta + \phi) + a \cos \beta$$
$$= 2a \cos \left(\frac{2\beta + \phi}{2} \right) \cos \left(\frac{\phi}{2} \right)$$
$$= 2a \cos \left(\frac{\phi}{2} \right) \cos \left(\beta + \frac{\phi}{2} \right).$$

The amplitude is $2a \cos(\phi/2)$, and the cosine wave is $\cos(\beta + \phi/2)$.

Finally, a general solution for two waves with the same frequency, but different amplitudes and phase angles:

$$\psi_1 = a_1 \cos(\beta + \phi_1)$$
$$\psi_2 = a_2 \cos(\beta + \phi_2).$$

The aim is to create a wave equation of the form $A \cos \Phi$, where A is the amplitude, and references the original amplitudes a_1 & a_2 and phase angles ϕ_1 & ϕ_2, and Φ references the original frequency and phase angles:

$$\begin{aligned}
\psi_1 + \psi_2 &= a_1 \cos(\beta + \phi_1) + a_2 \cos(\beta + \phi_2) \\
&= a_1 \cos \beta \cos \phi_1 - a_1 \sin \beta \sin \phi_1 + a_2 \cos \beta \cos \phi_2 - a_2 \sin \beta \sin \phi_2 \\
&= a_1 \cos \beta \cos \phi_1 + a_2 \cos \beta \cos \phi_2 - a_1 \sin \beta \sin \phi_1 - a_2 \sin \beta \sin \phi_2 \\
&= \cos \beta (a_1 \cos \phi_1 + a_2 \cos \phi_2) - \sin \beta (a_1 \sin \phi_1 + a_2 \sin \phi_2). \quad (8.2)
\end{aligned}$$

At this point, (8.2) is of the form

$$\psi_1 + \psi_2 = X \cos \beta - Y \sin \beta$$

where

$$X = a_1 \cos \phi_1 + a_2 \cos \phi_2 \qquad (8.3)$$
$$Y = a_1 \sin \phi_1 + a_2 \sin \phi_2 \qquad (8.4)$$

and we have managed to separate the frequency terms from those containing the amplitudes and phase angles.

One can see that (8.3) and (8.4) possess a symmetry where X is the sum of two terms projected onto the x-axis, whilst Y is the sum of two terms projected onto the y-axis. Therefore, we can express X and Y in polar form as

$$X = r \cos \theta$$
$$Y = r \sin \theta$$

where $r^2 = X^2 + Y^2$ and θ is the angle associated with r:

$$
\begin{aligned}
r^2 &= (a_1 \cos \phi_1 + a_2 \cos \phi_2)^2 + (a_1 \sin \phi_1 + a_2 \sin \phi_2)^2 \\
&= a_1^2 \cos^2 \phi_1 + a_2^2 \cos^2 \phi_2 + 2a_1 a_2 \cos \phi_1 \cos \phi_2 \\
&\quad + a_1^2 \sin^2 \phi_1 + a_2^2 \sin^2 \phi_2 + 2a_1 a_2 \sin \phi_1 \sin \phi_2 \\
&= a_1^2 \left(\cos^2 \phi_1 + \sin^2 \phi_1 \right) + a_2^2 \left(\cos^2 \phi_2 + \sin^2 \phi_2 \right) \\
&\quad + 2a_1 a_2 (\cos \phi_1 \cos \phi_2 + \sin \phi_1 \sin \phi_2) \\
&= a_1^2 + a_2^2 + 2a_1 a_2 \cos(\phi_1 - \phi_2) \\
r &= \sqrt{a_1^2 + a_2^2 + 2a_1 a_2 \cos(\phi_1 - \phi_2)}.
\end{aligned}
$$

Next, $\tan \theta = Y/X$, therefore,

$$\theta = \tan^{-1} \left(\frac{a_1 \sin \phi_1 + a_2 \sin \phi_2}{a_1 \cos \phi_1 + a_2 \cos \phi_2} \right).$$

We now have

$$a_1 \cos \phi_1 + a_2 \cos \phi_2 = r \cos \theta$$
$$a_1 \sin \phi_1 + a_2 \sin \phi_2 = r \sin \theta$$

which when substituted in (8.2) give:

$$
\begin{aligned}
\psi_1 + \psi_2 &= \cos \beta (a \cos \phi_1 + b \cos \phi_2) - \sin \beta (a \sin \phi_1 + b \sin \phi_2) \\
&= \cos \beta (r \cos \theta) - \sin \beta (r \sin \theta) \\
&= r \cos(\beta + \theta)
\end{aligned}
$$

and can be written

$$\psi_1 + \psi_2 = \sqrt{a_1^2 + a_2^2 + 2a_1 a_2 \cos(\phi_1 - \phi_2)} \cdot \cos \left[\beta + \tan^{-1} \left[\frac{a_1 \sin \phi_1 + a_2 \sin \phi_2}{a_1 \cos \phi_1 + a_2 \cos \phi_2} \right) \right].$$
$$(8.5)$$

Solving the above equations using trigonometric identities is not always easy, and one can waste vast amounts of time trying to simplify expressions. So now let's see how complex exponentials help resolve the situation.

8.4   Using Complex Exponentials

The complex exponential notation $re^{i\theta}$ offers a very effective mathematical object for manipulating waves. As $re^{i\theta} = r\cos\theta + ir\sin\theta$, we can select the real part for cosine waves, or the imaginary part for sine waves.

Let's continue using $\beta = kx - \omega t$, and ϕ for the phase angle. Therefore, a wave equation is expressed

$$\Psi = ae^{i(\beta+\phi)}. \tag{8.6}$$

Equation (8.6) contains all the properties of a wave, either explicitly or implicitly: amplitude, wavelength, frequency, phase angle and phase speed.

Given two such waves

$$\Psi_1 = a_1 e^{i(\beta+\phi_1)}$$

$$\Psi_2 = a_2 e^{i(\beta+\phi_2)}$$

then a general solution is given by

$$\begin{aligned}
\Psi_1 + \Psi_2 &= a_1 e^{i\beta} e^{i\phi_1} + a_2 e^{i\beta} e^{i\phi_2} \\
&= e^{i\beta}\left(a_1 e^{i\phi_1} + a_2 e^{i\phi_2}\right) \\
&= e^{i\beta}(a_1\cos\phi_1 + ia_1\sin\phi_1 + a_2\cos\phi_2 + ia_2\sin\phi_2) \\
&= e^{i\beta}[(a_1\cos\phi_1 + a_2\cos\phi_2) + i(a_1\sin\phi_1 + a_2\sin\phi_2)] \\
&= e^{i\beta}\sqrt{(a_1\cos\phi_1 + a_2\cos\phi_2)^2 + (a_1\sin\phi_1 + a_2\sin\phi_2)^2} \cdot e^{i\tan^{-1}\left(\frac{a_1\sin\phi_1 + a_2\sin\phi_2}{a_1\cos\phi_1 + a_2\cos\phi_2}\right)}
\end{aligned}$$

$$\Psi_1 + \Psi_2 = \sqrt{(a_1\cos\phi_1 + a_2\cos\phi_2)^2 + (a_1\sin\phi_1 + a_2\sin\phi_2)^2} \cdot e^{i\left[\beta+\tan^{-1}\left(\frac{a_1\sin\phi_1 + a_2\sin\phi_2}{a_1\cos\phi_1 + a_2\cos\phi_2}\right)\right]} \tag{8.7}$$

Equation (8.7) is a general wave equation for different amplitudes and phase angles. The radical term can be simplified as before, which produces

$$\Psi_1 + \Psi_2 = \sqrt{a_1^2 + a_2^2 + 2a_1 a_2\cos(\phi_1 - \phi_2)} \cdot e^{i\left[\beta+\tan^{-1}\left(\frac{a_1\sin\phi_1 + a_2\sin\phi_2}{a_1\cos\phi_1 + a_2\cos\phi_2}\right)\right]} \tag{8.8}$$

We can either select the real or imaginary parts of (8.8), where the real part combines two cosine waves:

$$\psi_1 + \psi_2 = \sqrt{a_1^2 + a_2^2 + 2a_1 a_2\cos(\phi_1 - \phi_2)} \cdot \cos\left[\beta + \tan^{-1}\left(\frac{a_1\sin\phi_1 + a_2\sin\phi_2}{a_1\cos\phi_1 + a_2\cos\phi_2}\right)\right] \tag{8.9}$$

or the imaginary part, which combines two sine waves:

$$\psi_1 + \psi_2 = \sqrt{a_1^2 + a_2^2 + 2a_1 a_2 \cos(\phi_1 - \phi_2)} \cdot \sin\left[\beta + \tan^{-1}\left(\frac{a_1 \sin\phi_1 + a_2 \sin\phi_2}{a_1 \cos\phi_1 + a_2 \cos\phi_2}\right)\right]$$

$$(8.10)$$

Let's use (8.9) and (8.10) to provide expressions for various wave combinations. For the moment, we will assume that the waves are either both sine or cosine waves, and provide equations for both functions. I will illustrate each wave combination with graphs based on the cosine or the sine wave equation.

8.4.1 Same Frequency and Amplitude, but no Phase Angle

Combining two cosine waves with the same frequency and amplitude, but no phase angle:

$$\psi_1 = a \cos\beta$$
$$\psi_2 = a \cos\beta$$
$$\psi_1 + \psi_2 = 2a \cos\beta.$$

If ψ_1 and ψ_2 are sine waves, then

$$\psi_1 + \psi_2 = 2a \sin\beta.$$

Figure 8.2 shows graphs of the following cosine waveforms:

$$0 \leq t \leq 720°$$
$$\psi_1 = 4\cos t \quad \text{[blue]}$$
$$\psi_2 = 4\cos t \quad \text{[green]}$$
$$\psi_1 + \psi_2 = 8\cos t \quad \text{[red]}.$$

8.4.2 Same Frequency, Different Amplitudes, but no Phase Angle

Combining two cosine waves with the same frequency, different amplitudes, but no phase angle:

Fig. 8.2 Combining two cosine waves $\psi_1 = 4\cos t$ and $\psi_2 = 4\cos t$

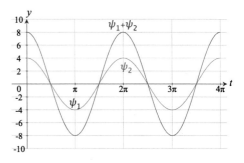

Fig. 8.3 Combining two sine waves $\psi_1 = 3\sin t$ and $\psi_2 = 5\sin t$

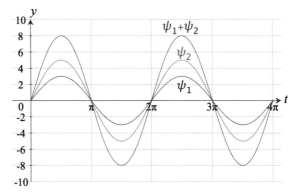

$$\psi_1 = a_1 \cos \beta$$
$$\psi_2 = a_2 \cos \beta$$
$$\psi_1 + \psi_2 = (a_1 + a_2) \cos \beta.$$

If ψ_1 and ψ_2 are sine waves, then

$$\psi_1 + \psi_2 = (a_1 + a_2) \sin \beta.$$

Figure 8.3 shows graphs of the following sine waveforms:

$$0 \le t \le 720°$$
$$\psi_1 = 3 \sin t \quad \text{[blue]}$$
$$\psi_2 = 5 \sin t \quad \text{[green]}$$
$$\psi_1 + \psi_2 = 8 \sin t \quad \text{[red]}.$$

Fig. 8.4 Combining two cosine waves $\psi_1 = 4\cos(t + 60°)$ and $4\cos(t + 60°)$

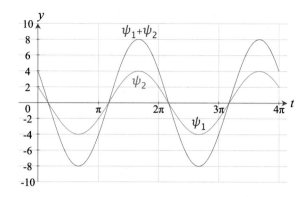

8.4.3 Same Frequency, Amplitude and Phase Angle

Combining two cosine waves with the same frequency, amplitude and phase angle:

$$\psi_1 = a\cos(\beta + \phi)$$
$$\psi_2 = a\cos(\beta + \phi)$$
$$\psi_1 + \psi_2 = 2a\cos(\beta + \phi).$$

If ψ_1 and ψ_2 are sine waves, then

$$\psi_1 + \psi_2 = 2a\sin(\beta + \phi).$$

Figure 8.4 shows graphs of the following cosine waveforms:

$$0 \le t \le 720°$$
$$\psi_1 = 4\cos(t + 60°) \quad \text{[blue]}$$
$$\psi_2 = 4\cos(t + 60°) \quad \text{[green]}$$
$$\psi_1 + \psi_2 = 8\cos(t + 60°) \quad \text{[red].}$$

8.4.4 Same Frequency and Amplitude, but Different Phase Angles

Combining two cosine waves with the same frequency and amplitude, but different phase angles:

Fig. 8.5 Combining two
sine waves
$\psi_1 = 4\sin(t + 60°)$ and
$\psi_2 = 4\sin(t + 30°)$

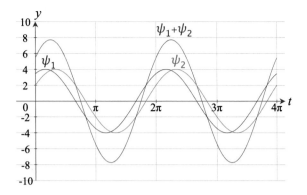

$$\psi_1 = a\cos(\beta + \phi_1)$$
$$\psi_2 = a\cos(\beta + \phi_2)$$
$$\psi_1 + \psi_2 = a\sqrt{2(1 + \cos(\phi_1 - \phi_2))} \cdot \cos\left[\beta + \tan^{-1}\left(\frac{\sin\phi_1 + \sin\phi_2}{\cos\phi_1 + \cos\phi_2}\right)\right].$$

If ψ_1 and ψ_2 are sine waves, then

$$\psi_1 + \psi_2 = a\sqrt{2(1 + \cos(\phi_1 - \phi_2))} \cdot \sin\left[\beta + \tan^{-1}\left(\frac{\sin\phi_1 + \sin\phi_2}{\cos\phi_1 + \cos\phi_2}\right)\right].$$

Figure 8.5 shows graphs of the following sine waveforms:

$$0 \le t \le 720°$$
$$\psi_1 = 4\sin(t + 60°) \quad \text{[blue]}$$
$$\psi_2 = 4\sin(t + 30°) \quad \text{[green]}$$
$$\psi_1 + \psi_2 \approx 7.727\sin(t + 45°) \quad \text{[red]}.$$

8.4.5 Same Frequency and Amplitude, but One has a Phase Angle

Combining two cosine waves with the same frequency and amplitude, but one
includes a phase angle:

Fig. 8.6 Combining two cosine waves $\psi_1 = 4\cos(t + 60°)$ and $\psi_2 = 4\cos t$

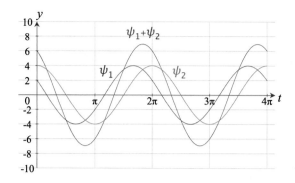

$$\psi_1 = a\cos(\beta + \phi)$$
$$\psi_2 = a\cos\beta$$
$$\psi_1 + \psi_2 = \sqrt{a^2 + a^2 + 2a^2\cos\phi} \cdot \cos\left[\beta + \tan^{-1}\left(\frac{a\sin\phi}{a\cos\phi + a}\right)\right]$$
$$= a\sqrt{2 + 2\cos\phi} \cdot \cos\left[\beta + \tan^{-1}\left(\frac{2\sin(\phi/2)\cos(\phi/2)}{2 - 2\sin^2(\phi/2)}\right)\right]$$
$$= a\sqrt{2 + 2 - 4\sin^2(\phi/2)} \cdot \cos\left[\beta + \tan^{-1}\left(\frac{\sin(\phi/2)\cos(\phi/2)}{\cos^2(\phi/2)}\right)\right]$$
$$\psi_1 + \psi_2 = 2a\cos(\phi/2)\cos(\beta + \phi/2).$$

If ψ_1 and ψ_2 are sine waves, then

$$\psi_1 + \psi_2 = 2a\cos(\phi/2)\sin(\beta + \phi/2).$$

Figure 8.6 shows graphs of the following cosine waveforms:

$$0 \leq t \leq 720°$$
$$\psi_1 = 4\cos(t + 60°) \quad \text{[blue]}$$
$$\psi_2 = 4\cos t \quad \text{[green]}$$
$$\psi_1 + \psi_2 = 8\cos 30° \cos(t + 30°) \quad \text{[red]}.$$

8.4.6 Same Frequency, Different Amplitudes, and One has a Phase Angle

Combining two cosine waves with the same frequency, different amplitudes, and one includes a phase angle:

Fig. 8.7 Combining two
cosine waves
$\psi_1 = 3\cos(t + 60°)$ and
$\psi_2 = 5\cos t$

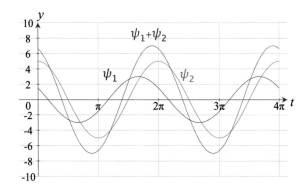

$$\psi_1 = a_1 \cos(\beta + \phi)$$

$$\psi_2 = a_2 \cos \beta$$

$$\psi_1 + \psi_2 = \sqrt{a_1^2 + a_2^2 + 2a_1 a_2 \cos \phi} \cdot \cos\left[\beta + \tan^{-1}\left(\frac{a_1 \sin \phi}{a_1 \cos \phi + a_2}\right)\right].$$

If ψ_1 and ψ_2 are sine waves, then

$$\psi_1 + \psi_2 = \sqrt{a_1^2 + a_2^2 + 2a_1 a_2 \cos \phi} \cdot \sin\left[\beta + \tan^{-1}\left(\frac{a_1 \sin \phi}{a_1 \cos \phi + a_2}\right)\right].$$

Figure 8.7 shows graphs of the following cosine waveforms:

$$0 \le t \le 720°$$

$$\psi_1 = 3\cos(t + 60°) \quad \text{[blue]}$$

$$\psi_2 = 5\cos t \quad \text{[green]}$$

$$\psi_1 + \psi_2 = \sqrt{3^2 + 5^2 + 30\cos 60°} \cdot \cos\left[t + \tan^{-1}\left(\frac{3\sin 60°}{3\cos 60° + 5}\right)\right]$$

$$\approx 7\cos(t + 18.875°) \quad \text{[red]}.$$

8.4.7   Same Frequency and Phase Angle, but Different Amplitudes

Combining two cosine waves with the same frequency and phase angle, but different amplitudes:

Fig. 8.8 Combining two sine waves $\psi_1 = 3\sin(t + 60°)$ and $\psi_2 = 5\sin(t + 60°)$

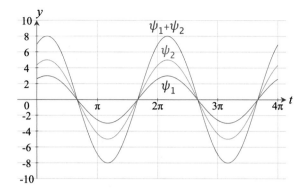

$$\psi_1 = a_1 \cos(\beta + \phi)$$
$$\psi_2 = a_2 \cos(\beta + \phi)$$
$$\psi_1 + \psi_2 = (a_1 + a_2)\cos(\beta + \phi).$$

If ψ_1 and ψ_2 are sine waves, then

$$\psi_1 + \psi_2 = (a_1 + a_2)\sin(\beta + \phi).$$

Figure 8.8 shows graphs of the following sine waveforms:

$$0 \leq t \leq 720°$$
$$\psi_1 = 3\sin(t + 60°) \quad \text{[blue]}$$
$$\psi_2 = 5\sin(t + 60°) \quad \text{[green]}$$
$$\psi_1 + \psi_2 = 8\sin(t + 60°) \quad \text{[red]}.$$

8.4.8 Same Frequency, but Different Amplitudes and Phase Angles

Combining two cosine waves with the same frequency, but different amplitudes and phase angles:

$$\psi_1 = a_1 \cos(\beta + \phi_1)$$
$$\psi_2 = a_2 \cos(\beta + \phi_2)$$
$$\psi_1 + \psi_2 = \sqrt{a_1^2 + a_2^2 + 2a_1 a_2 \cos(\phi_1 - \phi_2)} \cdot \cos\left[\beta + \tan^{-1}\left(\frac{a_1 \sin\phi_1 + a_2 \sin\phi_2}{a_1 \cos\phi_1 + a_2 \cos\phi_2}\right)\right].$$

Fig. 8.9 Combining two
cosine waves
$\psi_1 = 3\cos(t + 60°)$ and
$\psi_2 = 4\cos(t + 30°)$

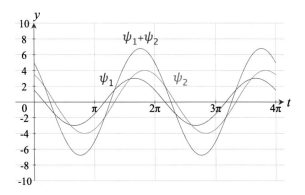

If ψ_1 and ψ_2 are sine waves, then

$$\psi_1 + \psi_2 = \sqrt{a_1^2 + a_2^2 + 2a_1a_2\cos(\phi_1 - \phi_2)} \cdot \sin\left[\beta + \tan^{-1}\left(\frac{a_1\sin\phi_1 + a_2\sin\phi_2}{a_1\cos\phi_1 + a_2\cos\phi_2}\right)\right].$$

Figure 8.9 shows graphs of the following cosine waveforms:

$$0 \le t \le 720°$$
$$\psi_1 = 3\cos(t + 60°)$$
$$\psi_2 = 4\cos(t + 30°)$$
$$\psi_1 + \psi_2 = r\cos(t + \theta)$$
$$r = \sqrt{3^2 + 4^2 + 2 \times 3 \times 4\cos(60° - 30°)}$$
$$= \sqrt{25 + 12\sqrt{3}}$$
$$\approx 6.766$$
$$\theta = \tan^{-1}\left(\frac{3\sin 60° + 4\sin 30°}{3\cos 60° + 4\cos 30°}\right)$$
$$\approx \tan^{-1}\left(\frac{4.598}{4.9641}\right)$$
$$\approx 42.8°$$
$$\psi_1 + \psi_2 \approx 6.766\cos(t + 42.8°).$$

8.4.9 Combining Sine and Cosine Functions

The above wave combinations assume that both waves have a sine or cosine basis.
So let's derive similar equations for a mixture of sine and cosine waves of the form

$$\psi_1 = a_1 \cos(\beta + \phi_1)$$
$$\psi_2 = a_2 \sin(\beta + \phi_2)$$

which can be written as

$$\psi_1 = a_1 \cos(\beta + \phi_1)$$
$$\psi_2 = a_2 \cos(\beta + \phi_2 - \pi/2)$$

and means that the general wave equation (8.7) can be used where ϕ_2 is replaced by $\phi_2 - \pi/2$:

$$\Psi_1 + \Psi_2 = \sqrt{(a_1 \cos \phi_1 + a_2 \cos(\phi_2 - \pi/2))^2 + (a_1 \sin \phi_1 + a_2 \sin(\phi_2 - \pi/2))^2} \cdot$$
$$\exp\left\{i\left[\beta + \tan^{-1}\left(\frac{a_1 \sin \phi_1 + a_2 \sin(\phi_2 - \pi/2)}{a_1 \cos \phi_1 + a_2 \cos(\phi_2 - \pi/2)}\right)\right]\right\}. \qquad (8.11)$$

But

$$\cos(\phi_2 - \pi/2) = \sin \phi_2$$
$$\sin(\phi_2 - \pi/2) = -\cos \phi_2$$

which when substituted into (8.11) gives

$$\Psi_1 + \Psi_2 = \sqrt{(a_1 \cos \phi_1 + a_2 \sin \phi_2)^2 + (a_1 \sin \phi_1 - a_2 \cos \phi_2)^2} \cdot$$
$$\exp\left\{i\left[\beta + \tan^{-1}\left(\frac{a_1 \sin \phi_1 - a_2 \cos \phi_2}{a_1 \cos \phi_1 + a_2 \sin \phi_2}\right)\right]\right\}$$

whose real part gives

$$\psi_1 + \psi_2 = \sqrt{a_1^2 + a_2^2 - 2a_1 a_2 \sin(\phi_1 - \phi_2)} \cdot \cos\left[\beta + \tan^{-1}\left(\frac{a_1 \sin \phi_1 - a_2 \cos \phi_2}{a_1 \cos \phi_1 + a_2 \sin \phi_2}\right)\right].$$
$$(8.12)$$

Let's test (8.12) with some examples.

8.4.10 Same Frequency and Amplitude, but no Phase Angle

Combining a sine and cosine wave with the same frequency and amplitude, but no phase angle:

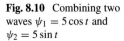

Fig. 8.10 Combining two
waves $\psi_1 = 5 \cos t$ and
$\psi_2 = 5 \sin t$

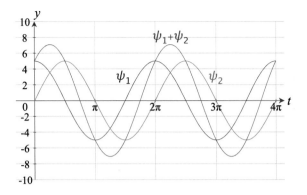

$$\psi_1 = a \cos \beta$$
$$\psi_2 = a \sin \beta$$
$$\psi_1 + \psi_2 = a\sqrt{2} \cos(\beta - \pi/4).$$

Figure 8.10 shows graphs of the following cosine and sine waveforms:

$$0 \leq t \leq 720°$$
$$\psi_1 = 5 \cos t \quad \text{[blue]}$$
$$\psi_2 = 5 \sin t \quad \text{[green]}$$
$$\psi_1 + \psi_2 = 5\sqrt{2} \cos(t - \pi/4) \quad \text{[red]}.$$

8.4.11 Same Frequency, Different Amplitudes, but no Phase Angle

Combining a sine and cosine wave with the same frequency, different amplitudes, but no phase angle:

$$\psi_1 = a_1 \cos \beta$$
$$\psi_2 = a_2 \sin \beta$$
$$\psi_1 + \psi_2 = \sqrt{a_1^2 + a_2^2} \cdot \cos\left[\beta - \tan^{-1}\left(\frac{a_2}{a_1}\right)\right].$$

Figure 8.11 shows graphs of the following cosine and sine waveforms:

Fig. 8.11 Combining two waves $\psi_1 = 4\cos t$ and $\psi_2 = 5\sin t$

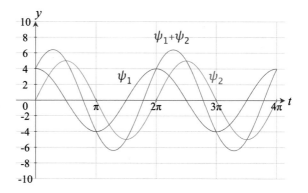

$$0 \le t \le 720°$$
$$\psi_1 = 4\cos t \quad \text{[blue]}$$
$$\psi_2 = 5\sin t \quad \text{[green]}$$
$$= \sqrt{41}\cos[t - \tan^{-1}(5/4)]$$
$$\psi_1 + \psi_2 \approx \sqrt{41}\cos(t - 51.35°) \quad \text{[red]}.$$

8.4.12 Same Frequency, Amplitude and Phase Angle

Combining a sine and cosine wave with the same frequency, amplitude and phase angle:

$$\psi_1 = a\cos(\beta + \phi)$$
$$\psi_2 = a\sin(\beta + \phi)$$
$$\psi_1 + \psi_2 = a\sqrt{2}\,\cos\left[\beta + \tan^{-1}\left(\frac{\sin\phi - \cos\phi}{\cos\phi + \sin\phi}\right)\right].$$

Figure 8.12 shows graphs of the following cosine and sine waveforms:

$$0 \le t \le 720°$$
$$\psi_1 = 5\cos(t + 60°) \quad \text{[blue]}$$
$$\psi_2 = 5\sin(t + 60°) \quad \text{[green]}$$
$$\psi_1 + \psi_2 \approx 7.071\cos(t + 15°) \quad \text{[red]}.$$

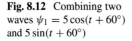

Fig. 8.12 Combining two
waves $\psi_1 = 5\cos(t + 60°)$
and $5\sin(t + 60°)$

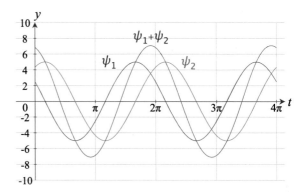

8.4.13 Same Frequency and Amplitude, but Different Phase Angles

Combining a sine and cosine wave with the same frequency and amplitude, but different phase angles:

$$\psi_1 = a\cos(\beta + \phi_1)$$
$$\psi_2 = a\sin(\beta + \phi_2)$$
$$\psi_1 + \psi_2 = a\sqrt{2(1 - \sin(\phi_1 - \phi_2))} \cdot \cos\left[\beta + \tan^{-1}\left(\frac{\sin\phi_1 - \cos\phi_2}{\cos\phi_1 + \sin\phi_2}\right)\right]$$

Figure 8.13 shows graphs of the following cosine and sine waveforms:

$$0 \le t \le 720°$$
$$\psi_1 = 5\cos(t + 60°) \quad \text{[blue]}$$
$$\psi_2 = 5\sin(t + 30°) \quad \text{[green]}$$
$$\psi_1 + \psi_2 = 5\cos t. \quad \text{[red]}$$

8.4.14 Same Frequency and Amplitude, but One has a Phase Angle

Combining a sine and cosine wave with the same frequency and amplitude, but one includes a phase angle:

Fig. 8.13 Combining two waves $\psi_1 = 5\cos(t + 60°)$ and $\psi_2 = 5\sin(t + 30°)$

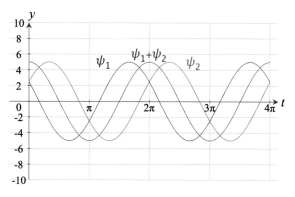

Fig. 8.14 Combining two waves $\psi_1 = 5\cos(t + 30°)$ and $\psi_2 = 5\sin t$

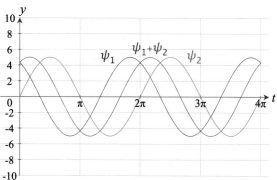

$$\psi_1 = a\cos(\beta + \phi)$$
$$\psi_2 = a\sin\beta$$
$$\psi_1 + \psi_2 = \sqrt{a^2 + a^2 - 2a^2\sin\phi} \cdot \cos\left[\beta - \tan^{-1}\left(\frac{1 - \sin\phi}{\cos\phi}\right)\right]$$
$$= a\sqrt{2(1 - \sin\phi)} \cdot \cos\left[\beta - \tan^{-1}\left(\frac{1 - \sin\phi}{\cos\phi}\right)\right].$$

Figure 8.14 shows graphs of the following cosine and sine waveforms:

$$0 \leq t \leq 720°$$
$$\psi_1 = 5\cos(t + 30°) \quad [\text{blue}]$$
$$\psi_2 = 5\sin t \quad [\text{green}]$$
$$\psi_1 + \psi_2 = 5\cos(t - 30°) \quad [\text{red}].$$

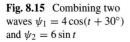

Fig. 8.15 Combining two waves $\psi_1 = 4\cos(t + 30°)$ and $\psi_2 = 6\sin t$

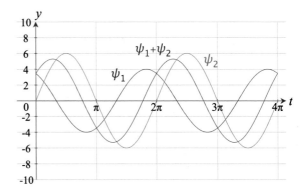

8.4.15 Same Frequency, Different Amplitudes, and One has a Phase Angle

Combining a sine and cosine wave with the same frequency, different amplitudes, and one includes a phase angle:

$$\psi_1 = a_1 \cos(\beta + \phi)$$
$$\psi_2 = a_2 \sin \beta$$
$$\psi_1 + \psi_2 = \sqrt{a_1^2 + a_2^2 - 2a_1a_2 \sin \phi} \cdot \cos \left[\beta + \tan^{-1} \left(\frac{a_1 \sin(\phi) - a_2}{a_1 \cos \phi} \right) \right]$$

Figure 8.15 shows graphs of the following cosine and sine waveforms:

$$0 \le t \le 720°$$
$$\psi_1 = 4\cos(t + 30°) \quad \text{[blue]}$$
$$\psi_2 = 6\sin t \quad \text{[green]}$$
$$\psi_1 + \psi_2 = \sqrt{4^2 + 6^2 - 48 \sin 30°} \cdot \cos \left[t + \tan^{-1} \left(\frac{4 \sin(30°) - 6}{4 \cos 30°} \right) \right]$$
$$\approx 5.292 \cos(t - 49.1°) \quad \text{[red]}.$$

8.4.16 Same Frequency, but Different Amplitudes and Phase Angles

Combining a sine and cosine wave with the same frequency, but different amplitudes and phase angles:

Fig. 8.16 Combining two waves $\psi_1 = 5\cos(t + 60°)$ and $\psi_2 = 6\sin(t + 30°)$

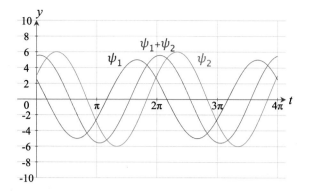

$$\psi_1 = a_1 \cos(\beta + \phi_1)$$

$$\psi_2 = a_2 \sin(\beta + \phi_2)$$

$$\psi_1 + \psi_2 = \sqrt{a_1^2 + a_2^2 - 2a_1 a_2 \sin(\phi_1 - \phi_2)} \cdot \cos\left[\beta + \tan^{-1}\left(\frac{a_1 \sin\phi_1 - a_2 \cos\phi_2}{a_1 \cos\phi_1 + a_2 \sin\phi_2}\right)\right].$$

Figure 8.16 shows graphs of the following cosine and sine waveforms:

$$0 \le t \le 720°$$

$$\psi_1 = 5\cos(t + 60°)$$

$$\psi_2 = 6\sin(t + 30°)$$

$$\psi_1 + \psi_2 = \sqrt{5^2 + 6^2 - 60\sin 30°} \cdot \cos\left[t + \tan^{-1}\left(\frac{5\sin 60° - 6\cos 30°}{5\cos 60° + 6\sin 30°}\right)\right]$$

$$\psi_1 + \psi_2 \approx 5.568 \cos(t - 8.914°).$$

8.4.17 Adding Several Cosine Waves

Now let's consider the problem of adding several, similar cosine waves together. For this exercise, let's assume that the waves share a common amplitude a, but have a constant phase difference δ:

$$\psi = a\cos(\beta) + a\cos(\beta + \delta) + a\cos(\beta + 2\delta) + a\cos(\beta + 3\delta). \qquad (8.13)$$

Equation (8.13) has a complex equivalent:

$$\Psi = ae^{i\beta} + ae^{i(\beta+\delta)} + ae^{i(\beta+2\delta)} + ae^{i(\beta+3\delta)}$$
$$= ae^{i\beta}\left(1 + e^{i\delta} + e^{i2\delta} + e^{i3\delta}\right)$$
$$= ae^{i\beta} Ae^{i\phi}$$

where

$$Ae^{i\phi} = 1 + e^{i\delta} + e^{i2\delta} + e^{i3\delta}$$

The problem to resolve is the value of A and ϕ. Boas and Arfken [1] provide an excellent solution in the form of the sum of a geometric series:

$$1 + r + r^2 + \cdots + r^{N-1} = \frac{1 - r^N}{1 - r}$$

therefore,

$$1 + e^{i\delta} + e^{i2\delta} + e^{i3\delta} = \frac{1 - e^{i4\delta}}{1 - e^{i\delta}}$$
$$= \frac{e^{i4\delta} - 1}{e^{i\delta} - 1}$$
$$= \frac{e^{i2\delta}}{e^{i\delta/2}} \frac{\left(e^{i2\delta} - e^{-i2\delta}\right)}{\left(e^{i\delta/2} - e^{-i\delta/2}\right)}$$
$$Ae^{i\phi} = \frac{\sin(2\delta)}{\sin(\delta/2)} e^{i3\delta/2}$$

therefore,

$$A = \frac{\sin(2\delta)}{\sin(\delta/2)}$$

and

$$\phi = 3\delta/2$$
$$\Psi = ae^{i\beta} \frac{\sin(2\delta)}{\sin(\delta/2)} e^{i3\delta/2}$$
$$= a \frac{\sin(2\delta)}{\sin(\delta/2)} e^{i(\beta+3\delta/2)}$$

consequently,

Fig. 8.17 Combining four cosine waves

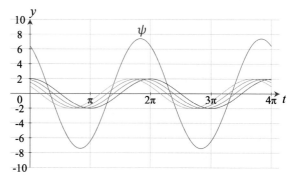

Fig. 8.18 Combining four sine waves

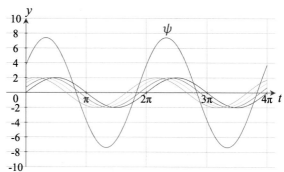

$$\psi = \text{Re}\left(a\frac{\sin(2\delta)}{\sin(\delta/2)}e^{i(\beta+3\delta/2)}\right)$$

$$= a\frac{\sin(2\delta)}{\sin(\delta/2)}\cos(\beta + 3\delta/2).$$

Figure 8.17 shows the combination of the following cosine waves

$$\psi = 2\cos t + 2\cos(t + 20°) + 2\cos(t + 40°) + 2\cos(t + 60°), \quad 0 \le t \le 720°.$$

If the waves are sine waves, then

$$\psi = \text{Im}\left(a\frac{\sin(2\delta)}{\sin(\delta/2)}e^{i(\beta+3\delta/2)}\right)$$

$$= a\frac{\sin(2\delta)}{\sin(\delta/2)}\sin(\beta + 3\delta/2).$$

Figure 8.18 shows the combination of the following sine waves

$$\psi = 2\sin t + 2\sin(t + 20°) + 2\sin(t + 40°) + 2\sin(t + 60°), \quad 0 \le t \le 720°.$$

Adding more waves is not a problem. For example, let's increase the number to five:

$$1 + e^{i\delta} + e^{i2\delta} + e^{i3\delta} + e^{i4\delta} = \frac{1 - e^{i5\delta}}{1 - e^{i\delta}}$$

$$= \frac{e^{i5\delta} - 1}{e^{i\delta} - 1}$$

$$= \frac{e^{i2.5\delta} \left(e^{i2.5\delta} - e^{-i2.5\delta}\right)}{e^{i\delta/2} \left(e^{i\delta/2} - e^{-i\delta/2}\right)}$$

$$Ae^{i\phi} = \frac{\sin(2.5\delta)}{\sin(\delta/2)} e^{i2\delta}$$

$$\Psi = ae^{i\beta} \frac{\sin(2.5\delta)}{\sin(\delta/2)} e^{i2\delta}$$

$$= a \frac{\sin(2.5\delta)}{\sin(\delta/2)} e^{i(\beta+2\delta)}$$

consequently,

$$\psi = \text{Re}\left(a \frac{\sin(2.5\delta)}{\sin(\delta/2)} e^{i(\beta+2\delta)}\right)$$

$$= a \frac{\sin(2.5\delta)}{\sin(\delta/2)} \cos(\beta + 2\delta).$$

Figure 8.19 shows the combination of the following cosine waves

$$\psi = 2\cos t + 2\cos(t + 20°) + 2\cos(t + 40°) + 2\cos(t + 60°) + 2\cos(t + 80°), \quad 0 \le t \le 720°.$$

One can now write a general equation for n cosines:

Fig. 8.19 Combining five cosine waves

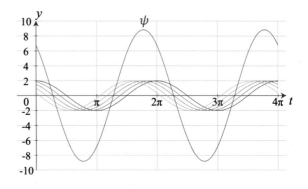

$$\text{four waves} \quad \psi = a\frac{\sin(2\delta)}{\sin(\delta/2)}\cos(\beta + 3\delta/2)$$

$$\text{five waves} \quad \psi = a\frac{\sin(2.5\delta)}{\sin(\delta/2)}\cos(\beta + 2\delta)$$

$$n \text{ waves} \quad \psi = a\frac{\sin(n\delta/2)}{\sin(\delta/2)}\cos(\beta + (n-1)\delta/2).$$

8.5 Summary

Hopefully, the above comparison between trigonometric identities and complex exponentials have convinced the reader the clarity imaginary mathematics brings to the area of wave combinations. The illustrations are included to bring a degree of reality to the whole subject.

Reference

1. Boas ML, Mathematical methods in the physical sciences. In: Arfken G (ed) Mathematical methods for physicists. http://www.physics.csbsju.edu/211/complex-review.pdf

Chapter 9
Circuit Analysis Using Complex Numbers

9.1 Introduction

In this chapter I show how complex numbers are used to resolve multi-phase currents in electrical circuits. I describe how a resistor, inductor and capacitor behave when subjected to an alternating voltage, and derive complex equations to express the instantaneous current, reactance and impedance.

9.2 Electronics

When studying electronics as a student, I came across the j operator, which is used to solve alternating currents in circuits containing resistors, capacitors and inductors. j is an imaginary unit: $j^2 = -1$, and is often used instead of i, which is used to represent current in electrical circuits. If you are not familiar with electrical circuits, here are some basic ideas.

9.2.1 Alternating Current and Voltages

Direct current (DC) is a stream of electrons moving in one direction, whilst with alternating current (AC), electrons move backwards and forwards, sinusoidally. A battery provides a constant voltage or potential difference for direct current, and a generator or electronic circuit produces an alternating voltage, to create an alternating current.

The SI unit for electrical potential difference is the volt [V], named after the Italian pioneer of electricity Alessandro Volta (1745–1827). The unit for current is

© Springer International Publishing AG, part of Springer Nature 2018
J. Vince, *Imaginary Mathematics for Computer Science*,
https://doi.org/10.1007/978-3-319-94637-5_9

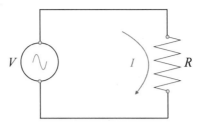

Fig. 9.1 A purely resistive circuit

the ampere [A], named after the French physicist and mathematician André-Marie Ampère (1775–1836).

9.2.2 Resistor

A resistor is an electrical component that impedes the flow of electrons, either DC or AC. The SI unit of *resistance R*, is the ohm [Ω], named after the German physicist Georg Simon Ohm (1789–1854). Figure 9.1 shows a pure resistive circuit, where resistance R is connected to the alternating voltage V.

For example, if $R = 100\,\Omega$, and $V = 12\,$V, the resulting current I is

$$I = \frac{V}{R} = \frac{12}{100} = 0.12\,\text{A} = 120\,\text{mA}.$$

This relationship is known as *Ohm's law*.

Figure 9.2 shows the voltage and current waveforms for a resistive circuit, which are always in phase.

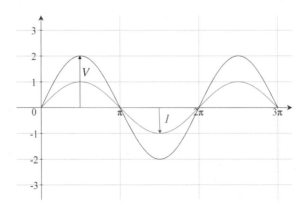

Fig. 9.2 Voltage and current waveforms in a pure resistive circuit

9.2.3 Inductor

An inductor is an electrical component in the form of a coil that opposes an alternating current. The SI unit of *inductance* L, is the henry [H], named after the American scientist Joseph Henry (1797–1878). Figure 9.3 shows a pure inductive circuit, where inductor L is connected to the alternating voltage V.

The inductance L of a circuit is the ratio of the induced voltage and the rate of change of the current, or

$$v(t) = L \frac{d}{dt} i(t)$$

where $v(t)$ is a time-varying voltage, and $i(t)$ a time-varying current. A coil with an inductance of 1 H generates a voltage 1 V across the coil when the current changes at 1 A/sec. Figure 9.4 shows the voltage waveform leading the current waveform by a phase angle of 90° for a perfect inductor.

An inductor's opposition to alternating current is called its *reactance* $X_L[\Omega]$, and is proportional to the current's frequency f, and inductance L; and because the voltage leads the current by 90°, is represented by the complex number $j\omega L$, where $\omega = 2\pi f$.

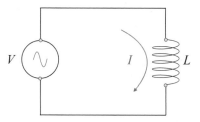

Fig. 9.3 A pure inductive circuit

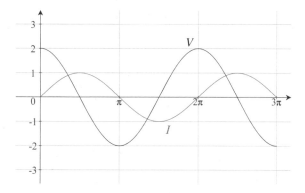

Fig. 9.4 Voltage and current waveforms in a perfect inductor

For example, if the alternating voltage has an amplitude of 10 V, and frequency 60 Hz, and the inductance $L = 40$ mH, then

$$
\begin{aligned}
I &= \frac{V}{X_L} = \frac{V}{j\omega L} \\
&= \frac{10}{j2\pi 60 \times 40 \times 10^{-3}} \\
&\approx \frac{10}{j15.08} \\
&\approx -j0.663 \text{ A}.
\end{aligned}
$$

The $-j$ implies that the current of 0.663 A lags the voltage by 90°.

9.2.4 Capacitor

A capacitor is an electrical component that stores energy in an electric field across two plates separated by an insulating dielectric. The SI unit of *capacitance* C is the farad [F], named after the English scientist Michael Faraday (1791–1867). Figure 9.5 shows a pure capacitive circuit, where capacitor C is connected to the alternating voltage V.

The capacitance C of a circuit is the ratio of the induced current and the rate of change of voltage, or

$$
i(t) = C\frac{d}{dt}v(t)
$$

where $v(t)$ is a time-varying voltage, and $i(t)$ a time-varying current. Figure 9.6 shows the voltage waveform trailing the current waveform by a phase angle of 90° for a pure capacitor.

A capacitor's opposition to alternating voltage is called its *reactance* $X_C[\Omega]$, and is inversely proportional to the current's frequency f, and capacitance C; and

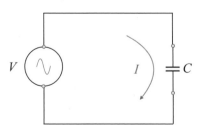

Fig. 9.5 A pure capacitive circuit

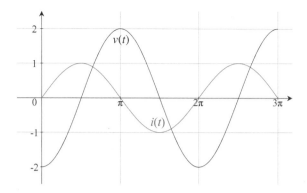

Fig. 9.6 Voltage and current waveforms in a perfect capacitor

because the voltage lags the current by 90°, it is represented by the complex number $1/j\omega C$.

For example, if the alternating voltage has an amplitude of 10 V, and frequency 60 Hz, and the capacitance $C = 60 \, \mu F$, then

$$I = \frac{V}{X_C} = jV\omega C$$
$$= j10 \times 2\pi 60 \times 60 \times 10^{-6}$$
$$\approx j0.226A.$$

The j implies that the current of 0.226 A leads the voltage by 90°.

9.2.5 Resistance, Reactance and Impedance

Resistance $[R]$ is the fixed opposition a circuit offers to the flow of electrons, irrespective of the voltage frequency. Reactance $[X]$ is the varying opposition a circuit offers to the flow of electrons, and depends on the voltage frequency. Impedance $[Z]$ is the combined opposition a resistive and reactive circuit offers to the flow of electrons, and depends on the voltage frequency. All three forms of electrical opposition are measured in units of ohms $[\Omega]$. Consequently,

$$Z = R + X_L + X_C$$
$$= R + j\omega L + \frac{1}{j\omega C}.$$

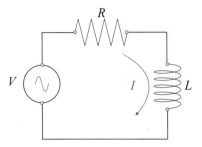

Fig. 9.7 A resistor and inductor circuit

Figure 9.7 shows a simple series circuit containing a resistor and an inductor. Let's calculate the current I, given $V = 10\,\text{V}, f = 50\,\text{Hz}, R = 50\,\Omega, L = 40\,\text{mH}$.

$$Z = R + j\omega L$$
$$= 50 + 2\pi 50 \times 40 \times 10^{-3} j$$
$$\approx 50 + 12.56 j$$
$$\approx 51.55 e^{i0.2461}$$
$$I = \frac{V}{Z}$$
$$\approx \frac{10}{51.55 e^{i0.2461}}$$
$$\approx 0.194 e^{-i0.2461}$$

where 0.194 A is the amplitude, and $e^{-i0.2461}$ is the frequency and phase of the complex current. The current waveform is the real part of I:

$$I(t) = \text{Re}\left(0.194 e^{-i0.2461}\right)$$
$$= 0.194 \cos(\omega t - 0.2461)$$
$$= 0.194 \cos(2\pi 50 t - 0.2461)$$
$$= 0.194 \cos(314 t - 0.2461)$$

Approximately, the current is 194 mA, and lags the voltage by 0.2461 rad, (14.1°).

Figure 9.8 shows a circuit with a resistor, inductor and capacitor in series with an alternating voltage source. Let's compute the current I, given $V = 10\,\text{V}, \ f = 50\,\text{Hz}, \ R = 50\,\Omega, \ L = 40\,\text{mH}, \ C = 50\,\mu\text{F}$.

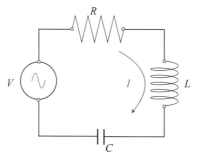

Fig. 9.8 A resistor, inductor and capacitor circuit

$$Z = R + j\omega L + 1/(j\omega C)$$

$$= 50 + 2\pi 50 \times 40 \times 10^{-3} j - \frac{j}{2\pi 50 \times 50 \times 10^{-6}}$$

$$\approx 50 + 12.566j - 63.662j$$

$$\approx 50 - 51.096j$$

$$\approx 71.49e^{-i0.7964}$$

$$I = \frac{V}{Z}$$

$$= \frac{10}{71.49e^{-i0.7964}}$$

$$= 0.14e^{i0.7964}$$

where 0.14 A is the amplitude, and $e^{i0.7964}$ is the frequency and phase of the complex current. The current waveform is the real part of I:

$$I(t) = \text{Re}\left(0.14e^{i0.7964}\right)$$

$$= 0.14\cos(\omega t + 0.7964)$$

$$= 0.14\cos(2\pi 50t + 0.7964)$$

$$= 0.14\cos(314t + 0.7964).$$

Approximately, the current is 140 mA, and leads the voltage by 0.7964 rad, (45.63°).

9.3 Summary

This chapter has shown how useful the imaginary operator j is in representing phase differences between current and voltage in simple electronic circuits. More elaborate circuits require larger systems of equations, and those of you who wish to study the subject further will discover many books and web sites.

Chapter 10
Geometry Using Geometric Algebra

10.1 Introduction

Traditionally, problems in geometry are solved by constructing "scaffolding" lines and some suitable vectors. Then, with the aid of the dot and cross product, we devise some parametric equations that hopefully reveal an answer. In this chapter I describe how geometric algebra concepts are used to solve some familiar problems in geometry. This approach relies upon the wedge product and pre- and post-multiplying by imaginary bivectors.

10.1.1 The Sine Rule

The sine rule states that for any triangle $\triangle ABC$ with angles α, β and θ, and respective opposite sides a, b and c, then

$$\frac{a}{\sin \alpha} = \frac{b}{\sin \beta} = \frac{c}{\sin \theta}.$$

This rule can be proved using the outer product of two vectors, which we know incorporates the sine of the angle between two vectors:

$$|\mathbf{a} \wedge \mathbf{b}| = |\mathbf{a}||\mathbf{b}| \sin \alpha.$$

With reference to Fig. 10.1 the area of $\triangle ABC$ can be expressed as

$$\text{area of } \triangle ABC = \tfrac{1}{2}|-\mathbf{c} \wedge \mathbf{a}| = \tfrac{1}{2}|\mathbf{c}||\mathbf{a}| \sin \beta$$
$$\text{area of } \triangle BCA = \tfrac{1}{2}|-\mathbf{a} \wedge \mathbf{b}| = \tfrac{1}{2}|\mathbf{a}||\mathbf{b}| \sin \theta$$
$$\text{area of } \triangle CAB = \tfrac{1}{2}|-\mathbf{b} \wedge \mathbf{c}| = \tfrac{1}{2}|\mathbf{b}||\mathbf{c}| \sin \alpha$$

© Springer International Publishing AG, part of Springer Nature 2018
J. Vince, *Imaginary Mathematics for Computer Science*,
https://doi.org/10.1007/978-3-319-94637-5_10

Fig. 10.1 The sine rule

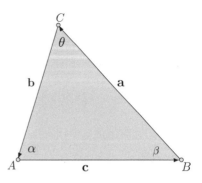

which means that

$$|\mathbf{c}||\mathbf{a}| \sin \beta = |\mathbf{a}||\mathbf{b}| \sin \theta = |\mathbf{b}||\mathbf{c}| \sin \alpha$$

therefore,

$$|\mathbf{a}| \sin \beta = |\mathbf{b}| \sin \alpha$$
$$|\mathbf{c}| \sin \beta = |\mathbf{b}| \sin \theta$$
$$\frac{|\mathbf{a}|}{\sin \alpha} = \frac{|\mathbf{b}|}{\sin \beta} = \frac{|\mathbf{c}|}{\sin \theta}.$$

10.1.2 The Cosine Rule

The cosine rule states that for any triangle $\triangle ABC$ with sides a, b and c, then

$$a^2 = b^2 + c^2 - 2bc \cos \alpha$$

where α is the angle between b and c.

Although this is an easy rule to prove using simple trigonometry, the geometric algebra solution is even easier.

Figure 10.2 shows a triangle $\triangle ABC$ constructed from vectors $\mathbf{a}$, $\mathbf{b}$ and $\mathbf{c}$. From Fig. 10.2

$$\mathbf{a} = \mathbf{b} - \mathbf{c}. \tag{10.1}$$

Squaring (10.1) we obtain

$$\mathbf{a}^2 = \mathbf{b}^2 + \mathbf{c}^2 - (\mathbf{bc} + \mathbf{cb}).$$

Fig. 10.2 The cosine rule

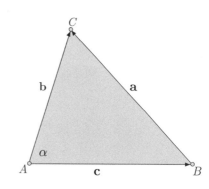

Fig. 10.3 A point P
perpendicular to a point T on
a line

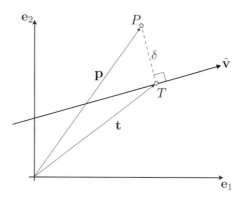

But

$$\mathbf{bc} + \mathbf{cb} = 2\mathbf{b} \cdot \mathbf{c} = 2|\mathbf{b}||\mathbf{c}| \cos \alpha$$

therefore,

$$|\mathbf{a}|^2 = |\mathbf{b}|^2 + |\mathbf{c}|^2 - 2|\mathbf{b}||\mathbf{c}| \cos \alpha.$$

10.1.3 A Point Perpendicular to a Line

Figure 10.3 shows a scenario where a line with direction vector $\hat{\mathbf{v}}$ passes through a point T. The objective is to locate another point P perpendicular to $\hat{\mathbf{v}}$ and a distance δ from T. The solution is found by post-multiplying $\hat{\mathbf{v}}$ by the imaginary pseudoscalar $\mathbf{e}_{12}$, which rotates $\hat{\mathbf{v}}$ through an angle of $90°$.

Fig. 10.4 A point P
perpendicular to a point T on
a line

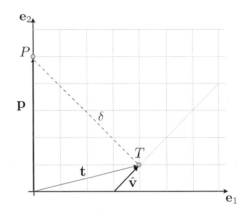

As $\hat{\mathbf{v}}$ is a unit vector

$$\overrightarrow{TP} = \delta\hat{\mathbf{v}}\mathbf{e}_{12}$$

therefore,

$$\mathbf{p} = \mathbf{t} + \overrightarrow{TP}$$

and

$$\mathbf{p} = \mathbf{t} + \delta\hat{\mathbf{v}}\mathbf{e}_{12}. \tag{10.2}$$

For example, Fig. 10.4 shows a 2D scenario where

$$\hat{\mathbf{v}} = \tfrac{1}{\sqrt{2}}(\mathbf{e}_1 + \mathbf{e}_2)$$
$$T = (4, 1)$$
$$\mathbf{t} = 4\mathbf{e}_1 + \mathbf{e}_2$$
$$\delta = \sqrt{32}.$$

Using (10.2)

$$\begin{aligned}
\mathbf{p} &= \mathbf{t} + \delta\hat{\mathbf{v}}\mathbf{e}_{12}\\
&= 4\mathbf{e}_1 + \mathbf{e}_2 + \sqrt{32}\tfrac{1}{\sqrt{2}}(\mathbf{e}_1 + \mathbf{e}_2)\mathbf{e}_{12}\\
&= 4\mathbf{e}_1 + \mathbf{e}_2 + 4\mathbf{e}_2 - 4\mathbf{e}_1\\
&= 5\mathbf{e}_2
\end{aligned}$$

and

$$P = (0, 5).$$

If $\mathbf{p}$ is required on the other side of the line, we pre-multiply $\hat{\mathbf{v}}$ by $\mathbf{e}_{12}$:

$$\mathbf{p} = \mathbf{t} + \delta\mathbf{e}_{12}\hat{\mathbf{v}}$$

which is the same as reversing the sign of δ.

10.1.4 Reflecting a Vector about a Vector

Reflecting a vector about another vector happens to be a rather easy problem for geometric algebra. Figure 10.5 shows the scenario where we see a vector $\mathbf{a}$ reflected about the normal to a line with direction vector $\hat{\mathbf{v}}$.

We begin by calculating $\hat{\mathbf{m}}$:

$$\hat{\mathbf{m}} = \hat{\mathbf{v}}\mathbf{e}_{12} \tag{10.3}$$

then reflecting $\mathbf{a}$ about $\hat{\mathbf{m}}$:

$$\mathbf{a}' = \hat{\mathbf{m}}\mathbf{a}\hat{\mathbf{m}}$$

substituting $\hat{\mathbf{m}}$ we have

$$\mathbf{a}' = \hat{\mathbf{v}}\mathbf{e}_{12}\mathbf{a}\hat{\mathbf{v}}\mathbf{e}_{12}. \tag{10.4}$$

Fig. 10.5 Reflecting a vector about a vector

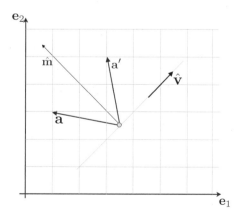

Fig. 10.6 Reflecting a vector about a vector

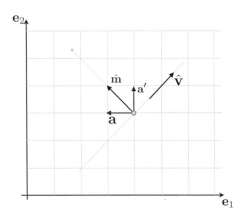

As an illustration, consider the scenario shown in Fig. 10.6 where

$$\hat{\mathbf{v}} = \tfrac{1}{\sqrt{2}}(\mathbf{e}_1 + \mathbf{e}_2)$$

$$\mathbf{a} = -\mathbf{e}_1$$

therefore, using (10.3)

$$\hat{\mathbf{m}} = \tfrac{1}{\sqrt{2}}(\mathbf{e}_1 + \mathbf{e}_2)\mathbf{e}_{12}$$

$$\hat{\mathbf{m}} = \tfrac{1}{\sqrt{2}}(\mathbf{e}_2 - \mathbf{e}_1)$$

and using (10.4)

$$\mathbf{a}' = \tfrac{1}{\sqrt{2}}(\mathbf{e}_2 - \mathbf{e}_1)(-\mathbf{e}_1)\tfrac{1}{\sqrt{2}}(\mathbf{e}_2 - \mathbf{e}_1)$$

$$= \tfrac{1}{2}(\mathbf{e}_{12} + 1)(\mathbf{e}_2 - \mathbf{e}_1)$$

$$= \tfrac{1}{2}(\mathbf{e}_1 + \mathbf{e}_2 + \mathbf{e}_2 - \mathbf{e}_1)$$

$$\mathbf{a}' = \mathbf{e}_2.$$

10.1.5   A Point Above or Below a Plane

In 3D geometry it is often required to test whether a point is above, below or on a planar surface. If we already have the plane equation for the surface it is just a question of substituting the test point in the equation and investigating its signed value. But here is another way using geometric algebra. For example, if a bivector is used to represent the orientation of a plane, the outer product of the test point's

Fig. 10.7 Point relative to a bivector

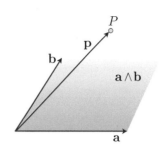

Fig. 10.8 Three points relative to a bivector

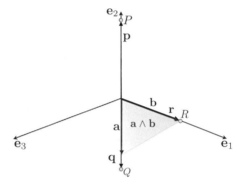

position vector with the bivector computes an oriented volume. Figure 10.7 shows a bivector $\mathbf{a} \wedge \mathbf{b}$ and a test point P with position vector $\mathbf{p}$ relative to the bivector.

Let

$\mathbf{a} \wedge \mathbf{b} \wedge \mathbf{p}$ is +ve, then P is "above" the bivector

$\mathbf{a} \wedge \mathbf{b} \wedge \mathbf{p}$ is −ve, then P is "below" the bivector

$\mathbf{a} \wedge \mathbf{b} \wedge \mathbf{p}$ is zero, then P is coplanar with the bivector.

The terms "above" and "below" mean in the bivector's positive and negative half-space respectively.

As an example, consider the scenario shown in Fig. 10.8 where the plane's orientation is represented by the bivector $\mathbf{a} \wedge \mathbf{b}$, and three test points P, Q and R.

If $P = (0, \ 1, \ 0), \quad Q = (0, \ -1, \ 0), \quad R = (1, \ 0, \ 0),$

$$\mathbf{a} = \mathbf{e}_1 + \mathbf{e}_3$$
$$\mathbf{b} = \mathbf{e}_1$$

then

$$\mathbf{p} = \mathbf{e}_2$$
$$\mathbf{q} = -\mathbf{e}_2$$
$$\mathbf{r} = \mathbf{e}_1$$

and

$$\mathbf{a} \wedge \mathbf{b} \wedge \mathbf{p} = (\mathbf{e}_1 + \mathbf{e}_3) \wedge \mathbf{e}_1 \wedge \mathbf{e}_2$$
$$= \mathbf{e}_{123}$$
$$\mathbf{a} \wedge \mathbf{b} \wedge \mathbf{q} = (\mathbf{e}_1 + \mathbf{e}_3) \wedge \mathbf{e}_1 \wedge (-\mathbf{e}_2)$$
$$= -\mathbf{e}_{123}$$
$$\mathbf{a} \wedge \mathbf{b} \wedge \mathbf{r} = (\mathbf{e}_1 + \mathbf{e}_3) \wedge \mathbf{e}_1 \wedge \mathbf{e}_1$$
$$= 0.$$

We can see that the signs of the first two volumes show that P is in the positive half-space, Q is in the negative half-space, and R is on the plane.

10.2   Summary

Hopefully, the above examples illustrate how useful geometric algebra is in resolving different types of geometric problems. And if you intend to discover more about the subject, then Chris Doran and Anthony Lasenby's book *Geometric Algebra for Physicists* is highly commended [1].

Reference

1. Doran C (2005) Geometric algebra for physicists. Cambridge University Press, Cambridge

Chapter 11
Rotating Vectors Using Quaternions

11.1 Introduction

In this chapter I show how quaternions are used to rotate vectors about an arbitrary axis. I begin by reviewing some of the history associated with quaternions, in particular, the role of Rodrigues, who discovered the importance of half-angles in rotation transforms.

For a particular quaternion product, when a quaternion is expressed as

$$q = [\cos\theta, \sin\theta\mathbf{v}]$$

a vector is rotated about the axis $\mathbf{v}$ by an angle θ. But, as we will discover, using a triple quaternion product, when a quaternion is expressed as

$$q = [\cos\left(\tfrac{\theta}{2}\right), \sin\left(\tfrac{\theta}{2}\right)\mathbf{v}]$$

a vector is rotated about the axis $\mathbf{v}$ by an angle θ. This half-angle representation was discovered by Rodrigues.

We examine various quaternion products to discover their rotational properties. This begins with two orthogonal quaternions, and moves towards the general case of using the triple qpq^{-1}, where q is a unit-norm quaternion, and p is a pure quaternion.

Two techniques are covered to express a quaternion product as a matrix, which in turn encode the eigenvector and eigenvalue. Finally, we examine how quaternions can be interpolated.

We continue to represent a quaternion as an ordered pair, with italic, lower-case letters to represent quaternions, and bold, lower-case letters to represent vectors.

© Springer International Publishing AG, part of Springer Nature 2018
J. Vince, *Imaginary Mathematics for Computer Science*,
https://doi.org/10.1007/978-3-319-94637-5_11

11.2 Some History

Rodrigues studied in Paris, and in 1816 was awarded his doctorate at the age of 21. The subject of his thesis was solving Legendre polynomials, and he proposed a solution which is still known as the *Rodrigues Formula*.

Although he pursued a career in politics and banking, his doctoral research confirms that he was more than just a "recreational" mathematician, for in 1840 he published a mathematical paper in the *Annales de Mathématiques Pures et Appliquées* on transformation groups [1]. The paper contains a formula describing a geometric construction equating two successive rotations about different axes, with a third rotation about another axis. Today, we know this correspondence as the *Euler–Rodrigues Parameterisation*. Euler had already shown in 1775 that a single rotation could represent two successive rotations about different axes, but did not provide an algebraic solution.

If we represent a rotation θ about an axial vector $\mathbf{v}$ as $\mathbf{R}_{\theta,\mathbf{v}}$, then Rodrigues provided the solution

$$\mathbf{R}_{\gamma,\mathbf{n}} = \mathbf{R}_{\alpha,\mathbf{l}}\mathbf{R}_{\beta,\mathbf{m}}$$

in the form of

$$\cos\left(\tfrac{\gamma}{2}\right) = \cos\left(\tfrac{\alpha}{2}\right)\cos\left(\tfrac{\beta}{2}\right) - \sin\left(\tfrac{\alpha}{2}\right)\sin\left(\tfrac{\beta}{2}\right)\mathbf{l}\cdot\mathbf{m} \qquad (11.1)$$

$$\sin\left(\tfrac{\gamma}{2}\right)\mathbf{n} = \sin\left(\tfrac{\alpha}{2}\right)\cos\left(\tfrac{\beta}{2}\right)\mathbf{l} + \cos\left(\tfrac{\alpha}{2}\right)\sin\left(\tfrac{\beta}{2}\right)\mathbf{m} + \sin\left(\tfrac{\alpha}{2}\right)\sin\left(\tfrac{\beta}{2}\right)\mathbf{l}\times\mathbf{m}. \qquad (11.2)$$

Rodrigues did not use the vector notation employed in (11.1) and (11.2), as this was yet to be defined by Hamilton, but he did employ the algebraic equivalent of these vector products. Figure 11.1 shows the spherical triangle formed by the axes and angles of rotation used by Rodrigues.

Fig. 11.1 Rodrigues' spherical triangle showing $\mathbf{l}$, $\mathbf{m}$ and $\mathbf{n}$

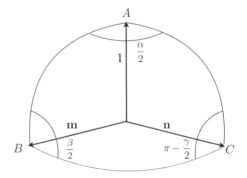

Equations (11.1) and (11.2) contain some features familiar to the quaternion product, which become obvious with the following analysis. We start by defining the quaternions

$$q_l = \left[\cos\left(\tfrac{\alpha}{2}\right), \; \sin\left(\tfrac{\alpha}{2}\right)\mathbf{l}\right]$$

$$q_m = \left[\cos\left(\tfrac{\beta}{2}\right), \; \sin\left(\tfrac{\beta}{2}\right)\mathbf{m}\right]$$

$$q_n = \left[\cos\left(\tfrac{\gamma}{2}\right), \; \sin\left(\tfrac{\gamma}{2}\right)\mathbf{n}\right]$$

and form the product

$$q_n = q_l q_m$$

$$= \left[\cos\left(\tfrac{\alpha}{2}\right), \; \sin\left(\tfrac{\alpha}{2}\right)\mathbf{l}\right]\left[\cos\left(\tfrac{\beta}{2}\right), \; \sin\left(\tfrac{\beta}{2}\right)\mathbf{m}\right]$$

$$= \left[\cos\left(\tfrac{\alpha}{2}\right)\cos\left(\tfrac{\beta}{2}\right) - \sin\left(\tfrac{\alpha}{2}\right)\sin\left(\tfrac{\beta}{2}\right)\mathbf{l}\cdot\mathbf{m},\right.$$

$$\left. \sin\left(\tfrac{\alpha}{2}\right)\cos\left(\tfrac{\beta}{2}\right)\mathbf{l} + \cos\left(\tfrac{\alpha}{2}\right)\sin\left(\tfrac{\beta}{2}\right)\mathbf{m} + \sin\left(\tfrac{\alpha}{2}\right)\sin\left(\tfrac{\beta}{2}\right)\mathbf{l}\times\mathbf{m}\right]$$

$$\cos\left(\tfrac{\gamma}{2}\right) = \cos\left(\tfrac{\alpha}{2}\right)\cos\left(\tfrac{\beta}{2}\right) - \sin\left(\tfrac{\alpha}{2}\right)\sin\left(\tfrac{\beta}{2}\right)\mathbf{l}\cdot\mathbf{m} \tag{11.3}$$

$$\sin\left(\tfrac{\gamma}{2}\right)\mathbf{n} = \sin\left(\tfrac{\alpha}{2}\right)\cos\left(\tfrac{\beta}{2}\right)\mathbf{l} + \cos\left(\tfrac{\alpha}{2}\right)\sin\left(\tfrac{\beta}{2}\right)\mathbf{m} + \sin\left(\tfrac{\alpha}{2}\right)\sin\left(\tfrac{\beta}{2}\right)\mathbf{l}\times\mathbf{m} \tag{11.4}$$

where (11.3) and (11.4) are identical to (11.1) and (11.2) respectively. Although Rodrigues had not invented quaternions in the form of

$$q = s + ai + bj + ck,$$

he had discovered the coefficients of a quaternion product before Hamilton. *C'est la vie!*

11.3 Quaternion Products

A quaternion q is the union of a scalar s and a vector $\mathbf{v}$:

$$q = [s, \; \mathbf{v}], \quad s \in \mathbb{R}, \quad \mathbf{v} \in \mathbb{R}^3.$$

If we express $\mathbf{v}$ in terms of its components, we have

$$q = [s, \; x\mathbf{i} + y\mathbf{j} + z\mathbf{k}], \quad \{s, x, y, z\} \in \mathbb{R}.$$

Hamilton had hoped that a quaternion could be used like a complex rotor, where

$$\mathbf{R}_\theta = \cos\theta + i\sin\theta$$

rotates a complex number by θ. Could a unit-norm quaternion q be used to rotate a vector stored as a pure quaternion p? Well yes, but only in a restricted sense. To understand this, let's construct the product of a unit-norm quaternion q and a pure quaternion p. The unit-norm quaternion q is defined as

$$q = [s,\ \lambda\hat{\mathbf{v}}], \quad \{s,\lambda\} \in \mathbb{R}, \quad \hat{\mathbf{v}} \in \mathbb{R}^3$$
$$|\hat{\mathbf{v}}| = 1$$
$$s^2 + \lambda^2 = 1$$

and the pure quaternion p stores the vector $\mathbf{p}$ to be rotated:

$$p = [0,\ \mathbf{p}], \quad \mathbf{p} \in \mathbb{R}^3.$$

Let's compute the product $p' = qp$ and examine the vector part of p' to see if $\mathbf{p}$ is rotated:

$$
\begin{aligned}
p' &= qp \\
&= [s,\ \lambda\hat{\mathbf{v}}][0,\ \mathbf{p}] \\
&= [-\lambda\hat{\mathbf{v}}\cdot\mathbf{p},\ s\mathbf{p} + \lambda\hat{\mathbf{v}}\times\mathbf{p}].
\end{aligned}
\tag{11.5}
$$

We can see from (11.5) that the result is a general quaternion with a scalar and a vector component.

11.3.1 Special Case

The "restricted sense" referred to above is that $\hat{\mathbf{v}}$ must be perpendicular to $\mathbf{p}$, which makes the dot product term $-\lambda\hat{\mathbf{v}}\cdot\mathbf{p}$ in (11.5) vanish, and we are left with the pure quaternion

$$p' = [0,\ s\mathbf{p} + \lambda\hat{\mathbf{v}}\times\mathbf{p}].
\tag{11.6}$$

Figure 11.2 illustrates this scenario, where $\mathbf{p}$ is perpendicular to $\hat{\mathbf{v}}$, and $\hat{\mathbf{v}}\times\mathbf{p}$ is perpendicular to the plane containing $\mathbf{p}$ and $\hat{\mathbf{v}}$. Now because $\hat{\mathbf{v}}$ is a unit vector, $|\mathbf{p}| = |\hat{\mathbf{v}}\times\mathbf{p}|$, which means that we have two orthogonal vectors: $\mathbf{p}$ and $\hat{\mathbf{v}}\times\mathbf{p}$, with the same length. Therefore, to rotate $\mathbf{p}$ about $\hat{\mathbf{v}}$, all that we have to do is make $s = \cos\theta$ and $\lambda = \sin\theta$ in (11.6):

Fig. 11.2 Three orthogonal
vectors $\mathbf{p}$, $\hat{\mathbf{v}}$ and $\hat{\mathbf{v}} \times \mathbf{p}$

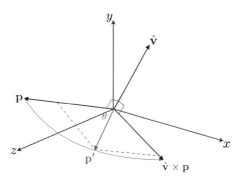

Fig. 11.3 The vector $\mathbf{p} = 2\mathbf{i}$
is rotated $45°$ by the
quaternion $q = \left[\frac{\sqrt{2}}{2}, \ \frac{\sqrt{2}}{2}\hat{\mathbf{v}} \right]$

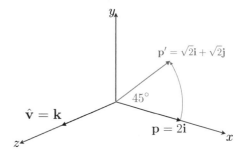

$$p' = [0, \ \mathbf{p}']$$
$$= [0, \ \cos\theta\mathbf{p} + \sin\theta\hat{\mathbf{v}} \times \mathbf{p}].$$

For example, to rotate a vector about the z-axis, q's vector $\hat{\mathbf{v}}$ must be aligned with the z-axis as shown in Fig. 11.3. If we make the angle of rotation $\theta = 45°$ then

$$q = [s, \ \lambda\hat{\mathbf{v}}]$$
$$= [\cos\theta, \ \sin\theta\mathbf{k}]$$
$$= \left[\frac{\sqrt{2}}{2}, \ \frac{\sqrt{2}}{2}\mathbf{k} \right]$$

and if the vector to be rotated is $\mathbf{p} = 2\mathbf{i}$, then

$$p = [0, \ \mathbf{p}]$$
$$= [0, \ 2\mathbf{i}].$$

There are now four product combinations worth exploring: qp, pq, $q^{-1}p$ and pq^{-1}. It's not worth considering qp^{-1} and $p^{-1}q$ as p^{-1} simply reverses the direction of $\mathbf{p}$. Let's start with qp:

$$p' = qp$$
$$= \left[\tfrac{\sqrt{2}}{2}, \ \tfrac{\sqrt{2}}{2}\mathbf{k} \right][0, \ 2\mathbf{i}]$$
$$= \left[0, \ 2\tfrac{\sqrt{2}}{2}\mathbf{i} + 2\tfrac{\sqrt{2}}{2}\mathbf{k} \times \mathbf{i} \right]$$
$$= \left[0, \ \sqrt{2}\mathbf{i} + \sqrt{2}\mathbf{j} \right]$$

and **p** has been rotated $45°$ to $\mathbf{p}' = \sqrt{2}\mathbf{i} + \sqrt{2}\mathbf{j}$.

Next, pq:

$$p' = pq$$
$$= [0, \ 2\mathbf{i}] \left[\tfrac{\sqrt{2}}{2}, \ \tfrac{\sqrt{2}}{2}\mathbf{k} \right]$$
$$= \left[0, \ 2\tfrac{\sqrt{2}}{2}\mathbf{i} - 2\tfrac{\sqrt{2}}{2}\mathbf{k} \times \mathbf{i} \right]$$
$$= \left[0, \ \sqrt{2}\mathbf{i} - \sqrt{2}\mathbf{j} \right]$$

and **p** has been rotated $-45°$ to $\mathbf{p}' = \sqrt{2}\mathbf{i} - \sqrt{2}\mathbf{j}$.

Next, $q^{-1}p$, and as q is a unit-norm quaternion, $q^{-1} = q^*$:

$$p' = q^{-1}p$$
$$= \left[\tfrac{\sqrt{2}}{2}, \ -\tfrac{\sqrt{2}}{2}\mathbf{k} \right][0, \ 2\mathbf{i}]$$
$$= \left[0, \ 2\tfrac{\sqrt{2}}{2}\mathbf{i} - 2\tfrac{\sqrt{2}}{2}\mathbf{k} \times \mathbf{i} \right]$$
$$= \left[0, \ \sqrt{2}\mathbf{i} - \sqrt{2}\mathbf{j} \right]$$

and **p** has been rotated $-45°$ to $\mathbf{p}' = \sqrt{2}\mathbf{i} - \sqrt{2}\mathbf{j}$.

Finally, pq^{-1}:

$$p' = pq^{-1}$$
$$= [0, \ 2\mathbf{i}] \left[\tfrac{\sqrt{2}}{2}, \ -\tfrac{\sqrt{2}}{2}\mathbf{k} \right]$$
$$= \left[0, \ 2\tfrac{\sqrt{2}}{2}\mathbf{i} + 2\tfrac{\sqrt{2}}{2}\mathbf{k} \times \mathbf{i} \right]$$
$$= \left[0, \ \sqrt{2}\mathbf{i} + \sqrt{2}\mathbf{j} \right]$$

and **p** has been rotated $45°$ to $\mathbf{p}' = \sqrt{2}\mathbf{i} + \sqrt{2}\mathbf{j}$. Thus, for orthogonal quaternions, θ is the angle of rotation, then

Fig. 11.4 Rotating the
vector $\mathbf{p} = 2\mathbf{i}$ by the
quaternion
$q = [\cos\theta,\ \sin\theta\hat{\mathbf{v}}]$

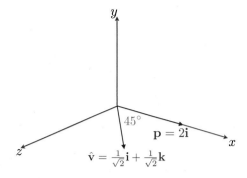

$$qp = pq^{-1}$$
$$pq = q^{-1}p.$$

Before moving on, let's see what happens to the product qp when $\theta = 180°$:

$$p' = qp$$
$$= [-1,\ \mathbf{0}][0,\ 2\mathbf{i}]$$
$$= [0,\ -2\mathbf{i}]$$

and $\mathbf{p}$ has been rotated $180°$ to $\mathbf{p}' = -2\mathbf{i}$.

Note that in all the above products, the vector has not been scaled during the rotation. This is because q is a unit-norm quaternion. Now let's see what happens if we change the angle between $\hat{\mathbf{v}}$ and $\mathbf{p}$. Let's reduce the angle to $45°$ and retain q's unit vector, as shown in Fig. 11.4. Therefore,

$$\hat{\mathbf{v}} = \tfrac{1}{\sqrt{2}}\mathbf{i} + \tfrac{1}{\sqrt{2}}\mathbf{k}$$
$$q = [\cos\theta,\ \sin\theta\hat{\mathbf{v}}]$$
$$p = [0,\ \mathbf{p}].$$

This time we must include the dot product term $-\sin\theta\hat{\mathbf{v}} \cdot \mathbf{p}$, as it is no longer zero:

$$p' = qp$$
$$= [\cos\theta,\ \sin\theta\hat{\mathbf{v}}][0,\ \mathbf{p}]$$
$$= [-\sin\theta\hat{\mathbf{v}} \cdot \mathbf{p},\ \cos\theta\mathbf{p} + \sin\theta\hat{\mathbf{v}} \times \mathbf{p}]. \qquad (11.7)$$

Substituting $\hat{\mathbf{v}}$, $\mathbf{p}$ and $\theta = 45°$ in (11.7), we have

$$p' = \left[-\frac{\sqrt{2}}{2}\left(\frac{1}{\sqrt{2}}\mathbf{i} + \frac{1}{\sqrt{2}}\mathbf{k}\right) \cdot (2\mathbf{i}), \ \frac{\sqrt{2}}{2}2\mathbf{i} + \frac{\sqrt{2}}{2}\left(\frac{1}{\sqrt{2}}\mathbf{i} + \frac{1}{\sqrt{2}}\mathbf{k}\right) \times 2\mathbf{i}\right]$$
$$= \left[-1, \ \sqrt{2}\mathbf{i} + \mathbf{j}\right] \tag{11.8}$$

which, unfortunately, is no longer a pure quaternion. It has not been rotated 45°, and the vector's norm is reduced to $\sqrt{3}$! Multiplying the vector by a non-orthogonal quaternion has converted some of the vector information into the quaternion's scalar component.

11.3.2 General Case

Not to worry. Could it be that an inverse quaternion reverses the operation? Let's see what happens if we post-multiply qp by q^{-1}.

Given

$$q = \left[\cos\theta, \ \sin\theta\left(\frac{1}{\sqrt{2}}\mathbf{i} + \frac{1}{\sqrt{2}}\mathbf{k}\right)\right]$$

then

$$q^{-1} = \left[\cos\theta, \ -\sin\theta\left(\frac{1}{\sqrt{2}}\mathbf{i} + \frac{1}{\sqrt{2}}\mathbf{k}\right)\right]$$
$$= \left[\frac{\sqrt{2}}{2}, \ \frac{-\sqrt{2}}{2}\left(\frac{1}{\sqrt{2}}\mathbf{i} + \frac{1}{\sqrt{2}}\mathbf{k}\right)\right]$$
$$= \frac{1}{2}\left[\sqrt{2}, \ -\mathbf{i} - \mathbf{k}\right].$$

Therefore, post-multiplying (11.8) by q^{-1} we have

$$qpq^{-1} = \left[-1, \ \sqrt{2}\mathbf{i} + \mathbf{j}\right]\frac{1}{2}\left[\sqrt{2}, \ -\mathbf{i} - \mathbf{k}\right]$$
$$= \frac{1}{2}\left[-\sqrt{2} - \left(\sqrt{2}\mathbf{i} + \mathbf{j}\right) \cdot (-\mathbf{i} - \mathbf{k}), \ \mathbf{i} + \mathbf{k} + \sqrt{2}\left(\sqrt{2}\mathbf{i} + \mathbf{j}\right) - \mathbf{i} + \sqrt{2}\mathbf{j} + \mathbf{k}\right]$$
$$= \frac{1}{2}\left[-\sqrt{2} + \sqrt{2}, \ \mathbf{i} + \mathbf{k} + 2\mathbf{i} + \sqrt{2}\mathbf{j} - \mathbf{i} + \sqrt{2}\mathbf{j} + \mathbf{k}\right]$$
$$= \left[0, \ \mathbf{i} + \sqrt{2}\mathbf{j} + \mathbf{k}\right] \tag{11.9}$$

which *is* a pure quaternion. Furthermore, there's no scaling as its norm is still 2, but the vector has been rotated 90° rather than 45°, twice the desired angle, as shown in Fig. 11.5.

If this "sandwiching" of the vector in the form of a pure quaternion by q and q^{-1} is correct, it suggests that increasing θ to 90° should rotate $\mathbf{p} = 2\mathbf{i}$ by 180° to $2\mathbf{k}$. Let's try this.

Fig. 11.5 The vector $\mathbf{p} = 2\mathbf{i}$
is rotated 90° to
$\mathbf{p}' = \mathbf{i} + \sqrt{2}\mathbf{j} + \mathbf{k}$

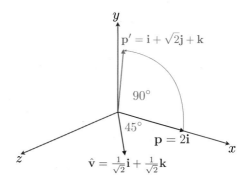

Let $\theta = 90°$, therefore,

$$qp = \left[0, \ \tfrac{1}{\sqrt{2}}\mathbf{i} + \tfrac{1}{\sqrt{2}}\mathbf{k}\right]\left[0, \ 2\mathbf{i}\right]$$
$$= \left[-\tfrac{2}{\sqrt{2}}, \ \tfrac{2}{\sqrt{2}}\mathbf{j}\right]$$

next, we post-multiply qp by q^{-1}:

$$qpq^{-1} = \left[-\tfrac{2}{\sqrt{2}}, \ \tfrac{2}{\sqrt{2}}\mathbf{j}\right]\left[0, \ -\tfrac{1}{\sqrt{2}}\mathbf{i} - \tfrac{1}{\sqrt{2}}\mathbf{k}\right]$$
$$= [0, \ \mathbf{i} + \mathbf{k} - \mathbf{i} + \mathbf{k}]$$
$$= [0, \ 2\mathbf{k}]$$

which confirms our prediction and suggests that qpq^{-1} works. Now let's show how this double angle arises. We begin by defining a unit-norm quaternion q:

$$q = [s, \ \lambda\hat{\mathbf{v}}]$$

where $s^2 + \lambda^2 = 1$.

The vector $\mathbf{p}$ to be rotated is encoded as a pure quaternion:

$$p = [0, \ \mathbf{p}]$$

and the inverse quaternion q^{-1} is

$$q^{-1} = [s, \ -\lambda\hat{\mathbf{v}}].$$

Therefore, the product qpq^{-1} is

$$qpq^{-1} = [s, \ \lambda\hat{\mathbf{v}}][0, \ \mathbf{p}][s, \ -\lambda\hat{\mathbf{v}}]$$

$$= \left[-\lambda\hat{\mathbf{v}} \cdot \mathbf{p}, \ s\mathbf{p} + \lambda\hat{\mathbf{v}} \times \mathbf{p}\right][s, \ -\lambda\hat{\mathbf{v}}]$$

$$= \left[-\lambda s\hat{\mathbf{v}} \cdot \mathbf{p} + \lambda s\mathbf{p} \cdot \hat{\mathbf{v}} + \lambda^2(\hat{\mathbf{v}} \times \mathbf{p}) \cdot \hat{\mathbf{v}},\right.$$

$$\left.\lambda^2(\hat{\mathbf{v}} \cdot \mathbf{p})\hat{\mathbf{v}} + s^2\mathbf{p} + \lambda s\hat{\mathbf{v}} \times \mathbf{p} - \lambda s\mathbf{p} \times \hat{\mathbf{v}} - \lambda^2(\hat{\mathbf{v}} \times \mathbf{p}) \times \hat{\mathbf{v}}\right]$$

$$= \left[\lambda^2(\hat{\mathbf{v}} \times \mathbf{p}) \cdot \hat{\mathbf{v}}, \ \lambda^2(\hat{\mathbf{v}} \cdot \mathbf{p})\hat{\mathbf{v}} + s^2\mathbf{p} + 2\lambda s\hat{\mathbf{v}} \times \mathbf{p} - \lambda^2(\hat{\mathbf{v}} \times \mathbf{p}) \times \hat{\mathbf{v}}\right].$$

Note that

$$(\hat{\mathbf{v}} \times \mathbf{p}) \cdot \hat{\mathbf{v}} = 0$$

and

$$(\hat{\mathbf{v}} \times \mathbf{p}) \times \hat{\mathbf{v}} = (\hat{\mathbf{v}} \cdot \hat{\mathbf{v}})\mathbf{p} - (\mathbf{p} \cdot \hat{\mathbf{v}})\hat{\mathbf{v}}$$

$$= \mathbf{p} - (\mathbf{p} \cdot \hat{\mathbf{v}})\hat{\mathbf{v}}.$$

Therefore,

$$qpq^{-1} = \left[0, \ \lambda^2\left(\hat{\mathbf{v}} \cdot \mathbf{p}\right)\hat{\mathbf{v}} + s^2\mathbf{p} + 2\lambda s\hat{\mathbf{v}} \times \mathbf{p} - \lambda^2\mathbf{p} + \lambda^2\left(\mathbf{p} \cdot \hat{\mathbf{v}}\right)\hat{\mathbf{v}}\right]$$

$$= \left[0, \ 2\lambda^2\left(\hat{\mathbf{v}} \cdot \mathbf{p}\right)\hat{\mathbf{v}} + (s^2 - \lambda^2)\mathbf{p} + 2\lambda s\hat{\mathbf{v}} \times \mathbf{p}\right]. \tag{11.10}$$

Obviously, this is a pure quaternion as the scalar component is zero. However, it is not obvious where the angle doubling comes from. But look what happens when we make $s = \cos\theta$ and $\lambda = \sin\theta$:

$$qpq^{-1} = [0, \ 2\sin^2\theta\left(\hat{\mathbf{v}} \cdot \mathbf{p}\right)\hat{\mathbf{v}} + (\cos^2\theta - \sin^2\theta)\mathbf{p} + 2\sin\theta\cos\theta\hat{\mathbf{v}} \times \mathbf{p}]$$

$$= \left[0, \ (1 - \cos(2\theta))\left(\hat{\mathbf{v}} \cdot \mathbf{p}\right)\hat{\mathbf{v}} + \cos(2\theta)\mathbf{p} + \sin(2\theta)\hat{\mathbf{v}} \times \mathbf{p}\right].$$

The double-angle trigonometric terms emerge! Now, if we want this product to actually rotate the vector by θ, then we must build this in from the outset by halving θ in q:

$$q = \left[\cos\left(\tfrac{\theta}{2}\right), \ \sin\left(\tfrac{\theta}{2}\right)\hat{\mathbf{v}}\right] \tag{11.11}$$

which makes

$$qpq^{-1} = \left[0, \ (1 - \cos\theta)\left(\hat{\mathbf{v}} \cdot \mathbf{p}\right)\hat{\mathbf{v}} + \cos\theta\mathbf{p} + \sin\theta\hat{\mathbf{v}} \times \mathbf{p}\right]. \tag{11.12}$$

The product qpq^{-1} was discovered by Hamilton who failed to publish the result. Cayley, also discovered the product and published the result in 1845 [2]. However, Altmann notes that "in Cayley's collected papers he concedes priority to Hamilton." [3], which was a nice gesture. However, the person who had recognised the importance of the half-angle parameters in (11.11) before Hamilton and Cayley was

Rodrigues – who published a solution that was not seen by Hamilton, but apparently, was seen by Cayley.

Let's test (11.12) using the previous example where we rotated a vector $\mathbf{p} = 2\mathbf{i}$, $\theta = 90°$ about the quaternion's vector $\hat{\mathbf{v}} = \frac{1}{\sqrt{2}}\mathbf{i} + \frac{1}{\sqrt{2}}\mathbf{k}$.

$$
\begin{aligned}
qpq^{-1} &= \left[0, \; (1-\cos\theta)(\hat{\mathbf{v}}\cdot\mathbf{p})\hat{\mathbf{v}} + \cos\theta\,\mathbf{p} + \sin\theta\,\hat{\mathbf{v}}\times\mathbf{p}\right] \\
&= \left[0, \; (\hat{\mathbf{v}}\cdot\mathbf{p})\hat{\mathbf{v}} + \hat{\mathbf{v}}\times\mathbf{p}\right] \\
&= \left[0, \; \tfrac{2}{\sqrt{2}}\left(\tfrac{1}{\sqrt{2}}\mathbf{i} + \tfrac{1}{\sqrt{2}}\mathbf{k}\right) + \sqrt{2}\mathbf{j}\right] \\
&= \left[0, \; \mathbf{i} + \sqrt{2}\mathbf{j} + \mathbf{k}\right]
\end{aligned}
$$

which agrees with (11.9). Thus, when a unit-norm quaternion takes the form

$$
q = \left[\cos\left(\tfrac{\theta}{2}\right), \; \sin\left(\tfrac{\theta}{2}\right)\hat{\mathbf{v}}\right]
$$

and a pure quaternion storing a vector to be rotated takes the form

$$
p = [0, \; \mathbf{p}]
$$

the pure quaternion

$$
p' = qpq^{-1}
$$

stores the rotated vector $\mathbf{p}'$. Let's show why this product preserves the magnitude of the rotated vector.

$$
\begin{aligned}
|p'| &= |qp||q^{-1}| \\
&= |q||p||q^{-1}| \\
&= |q|^2|p|
\end{aligned}
$$

and if q is a unit-norm quaternion, $|q| = 1$, then $|p'| = |p|$.

You may be wondering what happens if the product is reversed to $q^{-1}pq$? A guess would suggest that the rotation sequence is reversed, but let's see what an algebraic analysis confirms.

$$
\begin{aligned}
q^{-1}pq &= \left[s, \; -\lambda\hat{\mathbf{v}}\right]\left[0, \; \mathbf{p}\right]\left[s, \; \lambda\hat{\mathbf{v}}\right] \\
&= \left[\lambda\hat{\mathbf{v}}\cdot\mathbf{p}, \; s\mathbf{p} - \lambda\hat{\mathbf{v}}\times\mathbf{p}\right]\left[s, \; \lambda\hat{\mathbf{v}}\right] \\
&= \big[\lambda s\hat{\mathbf{v}}\cdot\mathbf{p} - \lambda s\mathbf{p}\cdot\hat{\mathbf{v}}, \\
&\qquad \lambda^2\hat{\mathbf{v}}\times\mathbf{p}\cdot\hat{\mathbf{v}} + \lambda^2\hat{\mathbf{v}}\cdot\mathbf{p}\hat{\mathbf{v}} + s^2\mathbf{p} - \lambda s\hat{\mathbf{v}}\times\mathbf{p} + \lambda s\mathbf{p}\times\hat{\mathbf{v}} - \lambda^2\hat{\mathbf{v}}\times\mathbf{p}\times\hat{\mathbf{v}}\big] \\
&= \left[\lambda^2(\hat{\mathbf{v}}\times\mathbf{p})\cdot\hat{\mathbf{v}}, \; \lambda^2(\hat{\mathbf{v}}\cdot\mathbf{p})\hat{\mathbf{v}} + s^2\mathbf{p} - 2\lambda s\hat{\mathbf{v}}\times\mathbf{p} - \lambda^2(\hat{\mathbf{v}}\times\mathbf{p})\times\hat{\mathbf{v}}\right].
\end{aligned}
$$

Once again

$$(\hat{\mathbf{v}} \times \mathbf{p}) \cdot \hat{\mathbf{v}} = 0$$

and

$$(\hat{\mathbf{v}} \times \mathbf{p}) \times \hat{\mathbf{v}} = \mathbf{p} - (\mathbf{p} \cdot \hat{\mathbf{v}})\hat{\mathbf{v}}.$$

Therefore,

$$
\begin{aligned}
q^{-1}pq &= \left[0,\; \lambda^2(\hat{\mathbf{v}} \cdot \mathbf{p})\hat{\mathbf{v}} + s^2\mathbf{p} - 2\lambda s\hat{\mathbf{v}} \times \mathbf{p} - \lambda^2\mathbf{p} + \lambda^2(\mathbf{p} \cdot \hat{\mathbf{v}})\hat{\mathbf{v}}\right] \\
&= \left[0,\; 2\lambda^2(\hat{\mathbf{v}} \cdot \mathbf{p})\hat{\mathbf{v}} + (s^2 - \lambda^2)\mathbf{p} - 2\lambda s\hat{\mathbf{v}} \times \mathbf{p}\right].
\end{aligned}
$$

Again, let's make $s = \cos\theta$ and $\lambda = \sin\theta$:

$$q^{-1}pq = \left[0,\; (1 - \cos(2\theta))(\hat{\mathbf{v}} \cdot \mathbf{p})\hat{\mathbf{v}} + \cos(2\theta)\mathbf{p} - \sin(2\theta)\hat{\mathbf{v}} \times \mathbf{p}\right]$$

and the only thing that has changed from qpq^{-1} is the sign of the cross-product term, which reverses the direction of its vector. However, we must remember to compensate for the angle-doubling by halving θ:

$$q^{-1}pq = \left[0,\; (1 - \cos\theta)(\hat{\mathbf{v}} \cdot \mathbf{p})\hat{\mathbf{v}} + \cos\theta\mathbf{p} - \sin\theta\hat{\mathbf{v}} \times \mathbf{p}\right]. \qquad (11.13)$$

Let's see what happens when we employ (11.13) to rotate $\mathbf{p} = 2\mathbf{i}$, $90°$ about the quaternion's vector $\hat{\mathbf{v}} = \frac{1}{\sqrt{2}}\mathbf{i} + \frac{1}{\sqrt{2}}\mathbf{k}$:

$$
\begin{aligned}
q^{-1}pq &= \left[0,\; \frac{2}{\sqrt{2}}\left(\frac{1}{\sqrt{2}}\mathbf{i} + \frac{1}{\sqrt{2}}\mathbf{k}\right) - \sqrt{2}\mathbf{j}\right] \\
&= [0,\; \mathbf{i} - \sqrt{2}\mathbf{j} + \mathbf{k}]
\end{aligned}
$$

which has rotated $\mathbf{p}$ clockwise $90°$ about the quaternion's vector. Therefore, the rotor qpq^{-1} rotates a vector counter-clockwise, and $q^{-1}pq$ rotates a vector clockwise:

$$
\begin{aligned}
qpq^{-1} &= \left[0,\; (1 - \cos\theta)(\hat{\mathbf{v}} \cdot \mathbf{p})\hat{\mathbf{v}} + \cos\theta\mathbf{p} + \sin\theta\hat{\mathbf{v}} \times \mathbf{p}\right] \\
q^{-1}pq &= \left[0,\; (1 - \cos\theta)(\hat{\mathbf{v}} \cdot \mathbf{p})\hat{\mathbf{v}} + \cos\theta\mathbf{p} - \sin\theta\hat{\mathbf{v}} \times \mathbf{p}\right].
\end{aligned}
$$

Let's compute another example. Consider the point $P(0, 1, 1)$ in Fig. 11.6 which is to be rotated $90°$ about the y-axis. We can see that the rotated point P' has the coordinates $(1, 1, 0)$ which we will confirm algebraically. The point P is represented by its position vector $\mathbf{p}$ in the pure quaternion

$$p = [0,\; \mathbf{p}].$$

The axis of rotation is $\hat{\mathbf{v}} = \mathbf{j}$, and the vector to be rotated is $\mathbf{p} = \mathbf{j} + \mathbf{k}$. Therefore,

Fig. 11.6 The point
$P(0, 1, 1)$ is rotated 90° to
$P'(1, 1, 0)$ about the y-axis

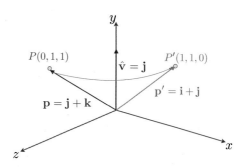

$$qpq^{-1} = \left[0,\ (1 - \cos\theta)(\hat{\mathbf{v}} \cdot \mathbf{p})\hat{\mathbf{v}} + \cos\theta\mathbf{p} + \sin\theta\hat{\mathbf{v}} \times \mathbf{p}\right]$$
$$= [0,\ \mathbf{j} \cdot (\mathbf{j} + \mathbf{k})\,\mathbf{j} + \mathbf{j} \times (\mathbf{j} + \mathbf{k})]$$
$$= [0,\ \mathbf{i} + \mathbf{j}]$$

and confirms that P is indeed rotated to $(1, 1, 0)$.

Now let's explore how this product is represented in matrix form.

11.4   Quaternions in Matrix Form

Having discovered a vector equation to represent the triple qpq^{-1}, let's continue and convert it into a matrix. We will explore two methods: the first is a simple vectorial method which translates the vector equation representing qpq^{-1} directly into matrix form. The second method uses matrix algebra to develop a rather cunning solution.

11.4.1   Vector Method

For the vector method it is convenient to describe the unit-norm quaternion as

$$q = [s,\ \mathbf{v}]$$
$$= [s,\ x\mathbf{i} + y\mathbf{j} + z\mathbf{k}]$$

where

$$s^2 + |\mathbf{v}|^2 = 1$$

and the pure quaternion as

$$p = [0, \ \mathbf{p}]$$
$$= [0, \ x_p\mathbf{i} + y_p\mathbf{j} + z_p\mathbf{k}].$$

A simple way to compute qpq^{-1} is to use (11.10) and substitute $|\mathbf{v}|$ for λ:

$$qpq^{-1} = \left[0, \ 2\lambda^2 \left(\hat{\mathbf{v}} \cdot \mathbf{p}\right) \hat{\mathbf{v}} + \left(s^2 - \lambda^2\right)\mathbf{p} + 2\lambda s\hat{\mathbf{v}} \times \mathbf{p}\right]$$
$$= \left[0, \ 2|\mathbf{v}|^2 \left(\hat{\mathbf{v}} \cdot \mathbf{p}\right) \hat{\mathbf{v}} + \left(s^2 - |\mathbf{v}|^2\right)\mathbf{p} + 2|\mathbf{v}|s\hat{\mathbf{v}} \times \mathbf{p}\right].$$

Next, we substitute $\mathbf{v}$ for $|\mathbf{v}|\hat{\mathbf{v}}$:

$$qpq^{-1} = \left[0, \ 2 \left(\mathbf{v} \cdot \mathbf{p}\right) \mathbf{v} + \left(s^2 - |\mathbf{v}|^2\right)\mathbf{p} + 2s\mathbf{v} \times \mathbf{p}\right].$$

Finally, as we are working with unit-norm quaternions to prevent scaling

$$s^2 + |\mathbf{v}|^2 = 1$$

and

$$s^2 - |\mathbf{v}|^2 = 2s^2 - 1$$

therefore,

$$qpq^{-1} = \left[0, \ 2(\mathbf{v} \cdot \mathbf{p})\mathbf{v} + \left(2s^2 - 1\right)\mathbf{p} + 2s\mathbf{v} \times \mathbf{p}\right].$$

If we let $p' = qpq^{-1}$, which is a pure quaternion, we have

$$p' = qpq^{-1}$$
$$= [0, \ \mathbf{p}']$$
$$= \left[0, \ 2(\mathbf{v} \cdot \mathbf{p})\mathbf{v} + \left(2s^2 - 1\right)\mathbf{p} + 2s\mathbf{v} \times \mathbf{p}\right]$$
$$\mathbf{p}' = 2(\mathbf{v} \cdot \mathbf{p})\mathbf{v} + \left(2s^2 - 1\right)\mathbf{p} + 2s\mathbf{v} \times \mathbf{p}.$$

We are only interested in the rotated vector $\mathbf{p}'$ comprising the three terms $2(\mathbf{v} \cdot \mathbf{p})\mathbf{v}$, $\left(2s^2 - 1\right)\mathbf{p}$ and $2s\mathbf{v} \times \mathbf{p}$, which can be represented by three individual matrices and summed together.

$$2(\mathbf{v} \cdot \mathbf{p})\mathbf{v} = 2\left(xx_p + yy_p + zz_p\right)\left(x\mathbf{i} + y\mathbf{j} + z\mathbf{k}\right)$$
$$= \begin{bmatrix} 2x^2 & 2xy & 2xz \\ 2xy & 2y^2 & 2yz \\ 2xz & 2yz & 2z^2 \end{bmatrix} \begin{bmatrix} x_p \\ y_p \\ z_p \end{bmatrix}$$
$$\left(2s^2 - 1\right)\mathbf{p} = \left(2s^2 - 1\right)x_p\mathbf{i} + \left(2s^2 - 1\right)y_p\mathbf{j} + \left(2s^2 - 1\right)z_p\mathbf{k}$$

$$= \begin{bmatrix} 2s^2 - 1 & 0 & 0 \\ 0 & 2s^2 - 1 & 0 \\ 0 & 0 & 2s^2 - 1 \end{bmatrix} \begin{bmatrix} x_p \\ y_p \\ z_p \end{bmatrix}$$

$$2s\mathbf{v} \times \mathbf{p} = 2s\Big((yz_p - zy_p)\mathbf{i} + (zx_p - xz_p)\mathbf{j} + (xy_p - yx_p)\mathbf{k} \Big)$$

$$= \begin{bmatrix} 0 & -2sz & 2sy \\ 2sz & 0 & -2sx \\ -2sy & 2sx & 0 \end{bmatrix} \begin{bmatrix} x_p \\ y_p \\ z_p \end{bmatrix}.$$

Adding these matrices together:

$$\mathbf{p}' = \begin{bmatrix} 2(s^2 + x^2) - 1 & 2(xy - sz) & 2(xz + sy) \\ 2(xy + sz) & 2(s^2 + y^2) - 1 & 2(yz - sx) \\ 2(xz - sy) & 2(yz + sx) & 2(s^2 + z^2) - 1 \end{bmatrix} \begin{bmatrix} x_p \\ y_p \\ z_p \end{bmatrix} \qquad (11.14)$$

or

$$\mathbf{p}' = \begin{bmatrix} 1 - 2(y^2 + z^2) & 2(xy - sz) & 2(xz + sy) \\ 2(xy + sz) & 1 - 2(x^2 + z^2) & 2(yz - sx) \\ 2(xz - sy) & 2(yz + sx) & 1 - 2(x^2 + y^2) \end{bmatrix} \begin{bmatrix} x_p \\ y_p \\ z_p \end{bmatrix} \qquad (11.15)$$

where

$$[0, \ \mathbf{p}'] = qpq^{-1}.$$

Now let's reverse the product. To compute the vector part of $q^{-1}pq$ all that we have to do is reverse the sign of $2s\mathbf{v} \times \mathbf{p}$:

$$\mathbf{p}' = \begin{bmatrix} 2(s^2 + x^2) - 1 & 2(xy + sz) & 2(xz - sy) \\ 2(xy - sz) & 2(s^2 + y^2) - 1 & 2(yz + sx) \\ 2(xz + sy) & 2(yz - sx) & 2(s^2 + z^2) - 1 \end{bmatrix} \begin{bmatrix} x_p \\ y_p \\ z_p \end{bmatrix} \qquad (11.16)$$

or

$$\mathbf{p}' = \begin{bmatrix} 1 - 2(y^2 + z^2) & 2(xy + sz) & 2(xz - sy) \\ 2(xy - sz) & 1 - 2(x^2 + z^2) & 2(yz + sx) \\ 2(xz + sy) & 2(yz - sx) & 1 - 2(x^2 + y^2) \end{bmatrix} \begin{bmatrix} x_p \\ y_p \\ z_p \end{bmatrix} \qquad (11.17)$$

where

$$[0, \; \mathbf{p}'] = q^{-1} p q.$$

Observe that (11.16) is the transpose of (11.14), and (11.17) is the transpose of (11.15).

11.4.2 Matrix Method

The second method to derive (11.12) employs the matrix representing a quaternion product:

$$q_a = [s_a, \; x_a\mathbf{i} + y_a\mathbf{j} + z_a\mathbf{k}]$$
$$q_b = [s_b, \; x_b\mathbf{i} + y_b\mathbf{j} + z_b\mathbf{k}]$$

and their product is

$$
\begin{aligned}
q_a q_b &= \left[s_a, \; x_a\mathbf{i} + y_a\mathbf{j} + z_a\mathbf{k} \right]\left[s_b, \; x_b\mathbf{i} + y_b\mathbf{j} + z_b\mathbf{k} \right]\\
&= \left[s_a s_b - x_a x_b - y_a y_b - z_a z_b, \right.\\
&\quad\; s_a\left(x_b\mathbf{i} + y_b\mathbf{j} + z_b\mathbf{k} \right) + s_b\left(x_a\mathbf{i} + y_a\mathbf{j} + z_a\mathbf{k} \right)\\
&\quad\; \left. + \left(y_a z_b - y_b z_a \right)\mathbf{i} + \left(x_b z_a - x_a z_b \right)\mathbf{j} + \left(x_a y_b - x_b y_a \right)\mathbf{k} \right]\\
&= \left[s_a s_b - x_a x_b - y_a y_b - z_a z_b, \right.\\
&\quad\; \left(s_a x_b + s_b x_a + y_a z_b - y_b z_a \right)\mathbf{i}\\
&\quad\; + \left(s_a y_b + s_b y_a + x_b z_a - x_a z_b \right)\mathbf{j}\\
&\quad\; \left. + \left(s_a z_b + s_b z_a + x_a y_b - x_b y_a \right)\mathbf{k} \right]\\
&= \begin{bmatrix} s_a & -x_a & -y_a & -z_a \\ x_a & s_a & -z_a & y_a \\ y_a & z_a & s_a & -x_a \\ z_a & -y_a & x_a & s_a \end{bmatrix} \begin{bmatrix} s_b \\ x_b \\ y_b \\ z_b \end{bmatrix} = \mathbf{A}q_b.
\end{aligned}
$$

At this stage we have quaternion q_a represented by matrix $\mathbf{A}$, and quaternion q_b represented as a column vector. Now let's reverse the scenario without altering the result by making q_b the matrix and q_a the column vector:

$$
q_a q_b = \begin{bmatrix} s_b & -x_b & -y_b & -z_b \\ x_b & s_b & z_b & -y_b \\ y_b & -z_b & s_b & x_b \\ z_b & y_b & -x_b & s_b \end{bmatrix} \begin{bmatrix} s_a \\ x_a \\ y_a \\ z_a \end{bmatrix} = \mathbf{B}q_a.
$$

So now we have two ways of computing $q_a q_b$ and we need a way of distinguishing between the two matrices. Let $\mathbf{L}$ be the matrix that preserves the left-to-right quaternion sequence, and $\mathbf{R}$ be the matrix that reverses the sequence to right-to-left:

$$q_a q_b = \mathbf{L}(q_a)q_b = \begin{bmatrix} s_a & -x_a & -y_a & -z_a \\ x_a & s_a & -z_a & y_a \\ y_a & z_a & s_a & -x_a \\ z_a & -y_a & x_a & s_a \end{bmatrix} \begin{bmatrix} s_b \\ x_b \\ y_b \\ z_b \end{bmatrix}$$

$$q_a q_b = \mathbf{R}(q_b)q_a = \begin{bmatrix} s_b & -x_b & -y_b & -z_b \\ x_b & s_b & z_b & -y_b \\ y_b & -z_b & s_b & x_b \\ z_b & y_b & -x_b & s_b \end{bmatrix} \begin{bmatrix} s_a \\ x_a \\ y_a \\ z_a \end{bmatrix}.$$

Remember that $\mathbf{L}(q_a)q_b = \mathbf{R}(q_b)q_a$, as this is central to understanding the next stage. Furthermore, don't be surprised if you can't follow the argument in the first reading. It took the author many hours of anguish trying to decipher the original algorithm, and this explanation has been expanded to ensure that you do not suffer the same experience!

First, let's employ the matrices $\mathbf{L}$ and $\mathbf{R}$ to rearrange the quaternion product $q_a q_c q_b$ to $q_a q_b q_c$. i.e. move q_c from the middle to the right-hand-side. We start with the quaternion product $q_a q_c q_b$ and divide it into two parts, $q_a q_c$ and q_b. We can do this because quaternion algebra is associative:

$$q_a q_c q_b = (q_a q_c)q_b.$$

We have already demonstrated above that the product $q_a q_c$ can be replaced by $\mathbf{L}(q_a)q_c$:

$$q_a q_c q_b = \mathbf{L}(q_a)q_c q_b.$$

We now have another two parts: $\mathbf{L}(q_a)q_c$ and q_b which can be reversed using $\mathbf{R}$ without disturbing the result:

$$q_a q_c q_b = \mathbf{L}(q_a)q_c q_b = \mathbf{R}(q_b)\mathbf{L}(q_a)q_c$$

which has achieved our objective to move q_c to the right-hand-side. But the most important result is that the matrices $\mathbf{R}(q_b)$ and $\mathbf{L}(q_a)$ can be multiplied together to form a single matrix, which operates on q_c.

Now let's repeat the same process to rearrange the product qpq^{-1}. The objective is to move p from the middle of q and q^{-1}, to the right-hand-side. The reason for doing this is to bring together q and q^{-1} in the form of two matrices, which can be multiplied together into a single matrix.

We start with the quaternion product qpq^{-1} and divide it into two parts, qp and q^{-1}:

$$qpq^{-1} = (qp)q^{-1}.$$

The product qp can be replaced by $\mathbf{L}(q)p$:

$$qpq^{-1} = \mathbf{L}(q)pq^{-1}.$$

We now have another two parts: $\mathbf{L}(q)p$ and q^{-1} which can be reversed using $\mathbf{R}$ without disturbing the result:

$$qpq^{-1} = \mathbf{L}(q)pq^{-1} = \mathbf{R}(q^{-1})\mathbf{L}(q)p$$

which has achieved our objective to move p to the right-hand-side.

The next step is to compute $\mathbf{L}(q)$ and $\mathbf{R}(q^{-1})$ using $q = [s, \ x\mathbf{i} + y\mathbf{j} + z\mathbf{k}]$. $\mathbf{L}(q)$ is easy as it is the same as $\mathbf{L}(q_a)$:

$$\mathbf{L}(q) = \begin{bmatrix} s & -x & -y & -z \\ x & s & -z & y \\ y & z & s & -x \\ z & -y & x & s \end{bmatrix}.$$

$\mathbf{R}(q^{-1})$ is also easy, but requires converting q_b in the original definition into q^{-1} which is effected by reversing the signs of the vector components:

$$\mathbf{R}(q^{-1}) = \begin{bmatrix} s & x & y & z \\ -x & s & -z & y \\ -y & z & s & -x \\ -z & -y & x & s \end{bmatrix}.$$

So now we can write

$$
\begin{aligned}
qpq^{-1} &= \mathbf{R}(q^{-1})\mathbf{L}(q)p \\[6pt]
&= \begin{bmatrix} s & x & y & z \\ -x & s & -z & y \\ -y & z & s & -x \\ -z & -y & x & s \end{bmatrix} \begin{bmatrix} s & -x & -y & -z \\ x & s & -z & y \\ y & z & s & -x \\ z & -y & x & s \end{bmatrix} \begin{bmatrix} 0 \\ x_p \\ y_p \\ z_p \end{bmatrix} \\[6pt]
&= \begin{bmatrix} 1 & 0 & 0 & 0 \\ 0 & 1-2(y^2+z^2) & 2(xy-sz) & 2(xz+sy) \\ 0 & 2(xy+sz) & 1-2(x^2+z^2) & 2(yz-sx) \\ 0 & 2(xz-sy) & 2(yz+sx) & 1-2(x^2+y^2) \end{bmatrix} \begin{bmatrix} 0 \\ x_p \\ y_p \\ z_p \end{bmatrix}.
\end{aligned}
$$

If we ignore the first row and column, the matrix computes $\mathbf{p}'$:

$$\mathbf{p}' = \begin{bmatrix} 1 - 2(y^2 + z^2) & 2(xy - sz) & 2(xz + sy) \\ 2(xy + sz) & 1 - 2(x^2 + z^2) & 2(yz - sx) \\ 2(xz - sy) & 2(yz + sx) & 1 - 2(x^2 + y^2) \end{bmatrix} \begin{bmatrix} x_p \\ y_p \\ z_p \end{bmatrix}$$

which is identical to (11.15)!

11.4.3 Geometric Verification

Let's illustrate the action of (11.14) by rotating the point $P(0, 1, 1)$, $90°$ about the y-axis, as shown in Fig. 11.6. The quaternion takes the form

$$q = \left[\cos \left(\tfrac{\theta}{2} \right), \ \sin \left(\tfrac{\theta}{2} \right) \hat{\mathbf{v}} \right]$$

which means that $\theta = 90°$ and $\hat{\mathbf{v}} = \mathbf{j}$, therefore,

$$q = \left[\cos 45°, \ \sin 45° \hat{\mathbf{j}} \right].$$

Consequently,

$$s = \tfrac{\sqrt{2}}{2}, \quad x = 0, \quad y = \tfrac{\sqrt{2}}{2}, \quad z = 0.$$

Substituting these values in (11.14) gives

$$\mathbf{p}' = \begin{bmatrix} 2(s^2 + x^2) - 1 & 2(xy - sz) & 2(xz + sy) \\ 2(xy + sz) & 2(s^2 + y^2) - 1 & 2(yz - sx) \\ 2(xz - sy) & 2(yz + sx) & 2(s^2 + z^2) - 1 \end{bmatrix} \begin{bmatrix} x_p \\ y_p \\ z_p \end{bmatrix}$$

$$\begin{bmatrix} 1 \\ 1 \\ 0 \end{bmatrix} = \begin{bmatrix} 0 & 0 & 1 \\ 0 & 1 & 0 \\ -1 & 0 & 0 \end{bmatrix} \begin{bmatrix} 0 \\ 1 \\ 1 \end{bmatrix}$$

where $P(0, 1, 1)$ is rotated to $P'(1, 1, 0)$, which is correct.

So now we have a transform that rotates a point about an arbitrary axis intersecting the origin, and can easily be implemented in software.

Before moving on, let's evaluate one more example. Let's perform a $180°$ rotation about a vector $\mathbf{v} = \mathbf{i} + \mathbf{k}$. To begin with, I will deliberately forget to convert the vector into a unit vector, just to see what happens to the final matrix. The quaternion takes the form

$$q = \left[\cos \left(\tfrac{\theta}{2} \right), \ \sin \left(\tfrac{\theta}{2} \right) \hat{\mathbf{v}} \right]$$

but we will use **v** as specified. Therefore, with $\theta = 180°$

$$s = 0, \quad x = 1, \quad y = 0, \quad z = 1.$$

Substituting these values in (11.14) gives

$$
\mathbf{p}' =
\begin{bmatrix}
2(s^2 + x^2) - 1 & 2(xy - sz) & 2(xz + sy) \\
2(xy + sz) & 2(s^2 + y^2) - 1 & 2(yz - sx) \\
2(xz - sy) & 2(yz + sx) & 2(s^2 + z^2) - 1
\end{bmatrix}
\begin{bmatrix}
x_p \\
y_p \\
z_p
\end{bmatrix}
$$

$$
=
\begin{bmatrix}
1 & 0 & 2 \\
0 & -1 & 0 \\
2 & 0 & 1
\end{bmatrix}
\begin{bmatrix}
1 \\
0 \\
0
\end{bmatrix}
$$

which looks nothing like a rotation matrix, and reminds us how important it is to have a unit vector to represent the axis. Let's repeat these calculations normalising the vector to $\hat{\mathbf{v}} = \frac{1}{\sqrt{2}}\mathbf{i} + \frac{1}{\sqrt{2}}\mathbf{k}$:

$$s = 0, \quad x = \frac{1}{\sqrt{2}}, \quad y = 0, \quad z = \frac{1}{\sqrt{2}}.$$

Substituting these values in (11.14) gives

$$
\mathbf{p}' =
\begin{bmatrix}
0 & 0 & 1 \\
0 & -1 & 0 \\
1 & 0 & 0
\end{bmatrix}
\begin{bmatrix}
1 \\
0 \\
0
\end{bmatrix}
$$

which not only looks like a rotation matrix, but has a determinant of 1 and rotates the point $P(1, 0, 0)$ to $P'(0, 0, 1)$ as shown in Fig. 11.7.

Fig. 11.7 The point
$P(1, 0, 0)$ is rotated $180°$
about the vector $\hat{\mathbf{v}}$ to
$P'(0, 0, 1)$

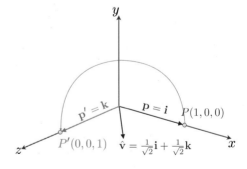

11.5 Multiple Rotations

Say a vector, or frame of reference, is subjected to two rotations specified by q_1 followed by q_2. There is a temptation to convert both quaternions to their respective matrix and multiply the matrices together. However, this not the most efficient way of combining the rotations. It is best to accumulate the rotations as quaternions and then convert to matrix notation, if required.

To illustrate this, consider the pure quaternion p subjected to the first quaternion q_1:

$$q_1 p q_1^{-1}$$

followed by a second quaternion q_2

$$q_2 \left(q_1 p q_1^{-1} \right) q_2^{-1}$$

which can be expressed as

$$(q_2 q_1) \, p \, (q_2 q_1)^{-1} .$$

Extra quaternions can be added accordingly. Let's illustrate this with two examples.

To keep things simple, the first quaternion q_1 rotates 30° about the y-axis:

$$q_1 = \left[\cos 15°, \ \sin 15° \mathbf{j} \right] .$$

The second quaternion q_2 rotates 60° also about the y-axis:

$$q_2 = \left[\cos 30°, \ \sin 30° \mathbf{j} \right] .$$

Together, the two quaternions rotate 90° about the y-axis. To accumulate these rotations, we multiply them together:

$$
\begin{aligned}
q_1 q_2 &= \left[\cos 15°, \ \sin 15° \mathbf{j} \right] \left[\cos 30°, \ \sin 30° \mathbf{j} \right] \\
&= \left[\cos 15° \cos 30° - \sin 15° \sin 30°, \ \cos 15° \sin 30° \mathbf{j} + \cos 30° \sin 15° \mathbf{j} \right] \\
&= \left[\tfrac{\sqrt{2}}{2}, \ \tfrac{\sqrt{2}}{2} \mathbf{j} \right]
\end{aligned}
$$

which is a quaternion that rotates 90° about the y-axis. Using the matrix (11.14) we have

$$s = \tfrac{\sqrt{2}}{2}, \quad x = 0, \quad y = \tfrac{\sqrt{2}}{2}, \quad z = 0,$$

$$
\begin{aligned}
\mathbf{p}' &= \begin{bmatrix} 2(s^2 + x^2) - 1 & 2(xy - sz) & 2(xz + sy) \\ 2(xy + sz) & 2(s^2 + y^2) - 1 & 2(yz - sx) \\ 2(xz - sy) & 2(yz + sx) & 2(s^2 + z^2) - 1 \end{bmatrix} \begin{bmatrix} x_p \\ y_p \\ z_p \end{bmatrix} \\
&= \begin{bmatrix} 0 & 0 & 1 \\ 0 & 1 & 0 \\ -1 & 0 & 0 \end{bmatrix} \begin{bmatrix} x_p \\ y_p \\ z_p \end{bmatrix}
\end{aligned}
$$

which rotates points about the y-axis by $90°$.

For a second example, let's just evaluate the quaternions. The first quaternion q_1 rotates $90°$ about the x-axis, and q_2 rotates $90°$ about the y-axis:

$$
q_1 = \left[\tfrac{\sqrt{2}}{2},\ \tfrac{\sqrt{2}}{2}\mathbf{i} \right]
$$

$$
q_2 = \left[\tfrac{\sqrt{2}}{2},\ \tfrac{\sqrt{2}}{2}\mathbf{j} \right]
$$

$$
p = [0,\ \mathbf{i} + \mathbf{j}]
$$

therefore,

$$
\begin{aligned}
q_2 q_1 &= \left[\tfrac{\sqrt{2}}{2}, \tfrac{\sqrt{2}}{2}\mathbf{j} \right]\left[\tfrac{\sqrt{2}}{2}, \tfrac{\sqrt{2}}{2}\mathbf{i} \right] \\
&= \left[\tfrac{1}{2}, \tfrac{\sqrt{2}}{2}\tfrac{\sqrt{2}}{2}\mathbf{i} + \tfrac{\sqrt{2}}{2}\tfrac{\sqrt{2}}{2}\mathbf{j} - \tfrac{1}{2}\mathbf{k} \right] \\
&= \left[\tfrac{1}{2}, \tfrac{1}{2}\mathbf{i} + \tfrac{1}{2}\mathbf{j} - \tfrac{1}{2}\mathbf{k} \right] \\
(q_2 q_1)^{-1} &= \left[\tfrac{1}{2}, -\tfrac{1}{2}\mathbf{i} - \tfrac{1}{2}\mathbf{j} + \tfrac{1}{2}\mathbf{k} \right] \\
(q_2 q_1)\, p &= \left[\tfrac{1}{2}, \tfrac{1}{2}\mathbf{i} + \tfrac{1}{2}\mathbf{j} - \tfrac{1}{2}\mathbf{k} \right][0,\ \mathbf{i} + \mathbf{j}] \\
&= \left[-\tfrac{1}{2} - \tfrac{1}{2}, \tfrac{1}{2}(\mathbf{i} + \mathbf{j}) + \tfrac{1}{2}\mathbf{i} - \tfrac{1}{2}\mathbf{j} \right] \\
&= [-1,\ \mathbf{i}] \\
(q_2 q_1)\, p\,(q_2 q_1)^{-1} &= [-1,\ \mathbf{i}]\left[\tfrac{1}{2}, -\tfrac{1}{2}\mathbf{i} - \tfrac{1}{2}\mathbf{j} + \tfrac{1}{2}\mathbf{k} \right] \\
&= \left[-\tfrac{1}{2} + \tfrac{1}{2}, \tfrac{1}{2}\mathbf{i} + \tfrac{1}{2}\mathbf{j} - \tfrac{1}{2}\mathbf{k} + \tfrac{1}{2}\mathbf{i} - \tfrac{1}{2}\mathbf{j} - \tfrac{1}{2}\mathbf{k} \right] \\
&= [0,\ \mathbf{i} - \mathbf{k}].
\end{aligned}
$$

Thus the point $(1, 1, 0)$ is rotated to $(1, 0, -1)$, which is correct.

11.6   Eigenvalue and Eigenvector

Although there is no doubt that (11.14) is a rotation matrix, we can secure further evidence by calculating its eigenvalue and eigenvector. The eigenvalue should be θ where

$$
\mathrm{Tr}\left(qpq^{-1}\right) = 1 + 2\cos\theta.
$$

and Tr is the trace function, which is the sum of the diagonal elements of a matrix.
The trace of (11.14) is

$$
\begin{aligned}
\mathrm{Tr}\left(qpq^{-1}\right) &= 2\left(s^2 + x^2\right) - 1 + 2\left(s^2 + y^2\right) - 1 + 2\left(s^2 + z^2\right) - 1 \\
&= 4s^2 + 2\left(s^2 + x^2 + y^2 + z^2\right) - 3 \\
&= 4s^2 - 1 \\
&= 4\cos^2\left(\tfrac{\theta}{2}\right) - 1 \\
&= 4\cos\theta + 4\sin^2\left(\tfrac{\theta}{2}\right) - 1 \\
&= 4\cos\theta + 2 - 2\cos\theta - 1 \\
&= 1 + 2\cos\theta
\end{aligned}
$$

and

$$
\cos\theta = \tfrac{1}{2}\left(\mathrm{Tr}\left(qpq^{-1}\right) - 1\right).
$$

To compute the eigenvector of (11.14) we use three equations derived as follows.

11.7 Analysis

We begin with the fact that a rotation matrix always has a real eigenvalue $\lambda = 1$,
which permits us to write

$$
\mathbf{Av} = \lambda\mathbf{v}
$$
$$
\mathbf{Av} = \lambda\mathbf{Iv} = \mathbf{Iv}
$$
$$
(\mathbf{A} - \mathbf{I})\,\mathbf{v} = \mathbf{0}
$$

therefore,

$$
\begin{bmatrix}
\left(a_{11} - 1\right) & a_{12} & a_{13} \\
a_{21} & \left(a_{22} - 1\right) & a_{23} \\
a_{31} & a_{32} & \left(a_{33} - 1\right)
\end{bmatrix}
\begin{bmatrix} x_v \\ y_v \\ z_v \end{bmatrix}
=
\begin{bmatrix} 0 \\ 0 \\ 0 \end{bmatrix}
\tag{11.18}
$$

Expanding (11.18) we have

$$
\begin{aligned}
\left(a_{11} - 1\right)x_v + a_{12}y_v + a_{13}z_v &= 0 \\
a_{21}x_v + \left(a_{22} - 1\right)y_v + a_{23}z_v &= 0 \\
a_{31}x_v + a_{32}y_v + \left(a_{33} - 1\right)z_v &= 0.
\end{aligned}
$$

There exists a trivial solution where $x_v = y_v = z_v = 0$, but to discover something more useful we can relax any one of the v terms which gives us three equations in two unknowns. Let's make $x_v = 0$:

$$a_{12}y_v + a_{13}z_v = -(a_{11} - 1) \tag{11.19}$$

$$(a_{22} - 1)y_v + a_{23}z_v = -a_{21} \tag{11.20}$$

$$a_{32}y_v + (a_{33} - 1)z_v = -a_{31}. \tag{11.21}$$

We are now faced with choosing a pair of equations to isolate y_v and z_v. In fact, we have to consider all three pairings because it is possible that a future rotation matrix will contain a column with two zero elements, which could conflict with any pairing we make at this stage.

Let's begin by choosing (11.19) and (11.20). The solution employs the following strategy: Given the following matrix equation

$$\begin{bmatrix} a_1 & b_1 \\ a_2 & b_2 \end{bmatrix} \begin{bmatrix} x \\ y \end{bmatrix} = \begin{bmatrix} c_1 \\ c_2 \end{bmatrix}$$

then

$$\frac{x}{\begin{vmatrix} c_1 & b_1 \\ c_2 & b_2 \end{vmatrix}} = \frac{y}{\begin{vmatrix} a_1 & c_1 \\ a_2 & c_2 \end{vmatrix}} = \frac{1}{\begin{vmatrix} a_1 & b_1 \\ a_2 & b_2 \end{vmatrix}}.$$

Therefore, using the 1st and 2nd Eqs. (11.19) and (11.20) we have

$$\frac{y_v}{\begin{vmatrix} -(a_{11} - 1) & a_{13} \\ -a_{21} & a_{23} \end{vmatrix}} = \frac{z_v}{\begin{vmatrix} a_{12} & -(a_{11} - 1) \\ (a_{22} - 1) & -a_{21} \end{vmatrix}} = \frac{1}{\begin{vmatrix} a_{12} & a_{13} \\ (a_{22} - 1) & a_{23} \end{vmatrix}}$$

$$x_v = a_{12}a_{23} - a_{13}(a_{22} - 1)$$
$$y_v = a_{13}a_{21} - a_{23}(a_{11} - 1)$$
$$z_v = (a_{11} - 1)(a_{22} - 1) - a_{12}a_{21}.$$

Similarly, using the 1st and 3rd Eqs. (11.19) and (11.21) we have

$$x_v = a_{12}(a_{33} - 1) - a_{13}a_{32}$$
$$y_v = a_{13}a_{31} - (a_{11} - 1)(a_{33} - 1)$$
$$z_v = a_{32}(a_{11} - 1) - a_{12}a_{31}$$

and using the 2nd and 3rd Eqs. (11.20) and (11.21) we have

$$x_v = (a_{22} - 1)(a_{33} - 1) - a_{23}a_{32}$$
$$y_v = a_{23}a_{31} - a_{21}(a_{33} - 1)$$
$$z_v = a_{21}a_{32} - a_{31}(a_{22} - 1).$$

Now we have nine equations to cope with any eventuality. In fact, there is nothing to stop us from choosing any three that take our fancy, for example these three equations look interesting and sound:

$$x_v = (a_{22} - 1)(a_{33} - 1) - a_{23}a_{32} \tag{11.22}$$
$$y_v = (a_{33} - 1)(a_{11} - 1) - a_{31}a_{13} \tag{11.23}$$
$$z_v = (a_{11} - 1)(a_{22} - 1) - a_{12}a_{21}. \tag{11.24}$$

Therefore, the solution for the eigenvector is $[x_v \ \ y_v \ \ z_v]^T$. Note that the sign of y_v has been reversed to maintain symmetry.

$$x_v = (a_{22} - 1)(a_{33} - 1) - a_{23}a_{32}$$
$$y_v = (a_{33} - 1)(a_{11} - 1) - a_{31}a_{13}$$
$$z_v = (a_{11} - 1)(a_{22} - 1) - a_{12}a_{21}.$$

Therefore,

$$\begin{aligned}
x_v &= \left(2(s^2 + y^2) - 2\right)\left(2(s^2 + z^2) - 2\right) - 2(yz - sx)2(yz + sx) \\
&= 4\left(s^2 + y^2 - 1\right)\left(s^2 + z^2 - 1\right) - 4\left(y^2z^2 - s^2x^2\right) \\
&= 4\left((x^2 + z^2)(x^2 + y^2) - y^2z^2 + s^2x^2\right) \\
&= 4\left(x^4 + x^2y^2 + x^2z^2 + z^2y^2 - y^2z^2 + s^2x^2\right) \\
&= 4x^2\left(s^2 + x^2 + y^2 + z^2\right) \\
&= 4x^2.
\end{aligned}$$

Similarly, $y_v = 4y^2$ and $z_v = 4z^2$, which confirm that the eigenvector has components associated with the quaternion's vector. The square terms should be no surprise, as the triple qpq^{-1} includes the product of three quaternions.

Let's test these formulae with the matrix associated with Fig. 11.7, which rotates a point $180°$ about the vector $\hat{\mathbf{v}} = \frac{1}{\sqrt{2}}\mathbf{i} + \frac{1}{\sqrt{2}}\mathbf{k}$:

$$\mathbf{M} = \begin{bmatrix} a_{11} & a_{12} & a_{13} \\ a_{21} & a_{22} & a_{23} \\ a_{31} & a_{32} & a_{33} \end{bmatrix} = \begin{bmatrix} 0 & 0 & 1 \\ 0 & -1 & 0 \\ 1 & 0 & 0 \end{bmatrix}$$

therefore,

$$x_v = -2 \times -1 - 0 = 2$$
$$y_v = -1 \times -1 - 1 \times 1 = 0$$
$$z_v = -1 \times -2 - 0 = 2$$

which confirms that the eigenvector is $2\mathbf{i} + 2\mathbf{k}$.

Next, $\text{Tr}(\mathbf{M}) = -1$, therefore

$$\cos \theta = \tfrac{1}{2}\left(\text{Tr}(qpq^{-1}) - 1\right)$$
$$= \tfrac{1}{2}\left((-1) - 1\right)$$
$$= -1$$
$$\theta = \pm 180°$$

which agrees with the previous results.

11.8 Interpolating Quaternions

Like vectors, quaternions can be interpolated to compute an in-between quaternion. However, whereas two interpolated vectors results in a third vector that is readily visualised, two interpolated quaternions results in a third quaternion that acts as a rotor, and is not immediately visualised.

The spherical interpolant for vectors is

$$\mathbf{v} = \frac{\sin[(1-t)\theta]}{\sin \theta}\mathbf{v}_1 + \frac{\sin(t\theta)}{\sin \theta}\mathbf{v}_2$$

where θ is the angle between the vectors, and requires no modification for quaternions:

$$q = \frac{\sin[(1-t)\theta]}{\sin \theta}q_1 + \frac{\sin(t\theta)}{\sin \theta}q_2. \tag{11.25}$$

So, given

$$q_1 = [s_1, \ x_1\mathbf{i} + y_1\mathbf{j} + z_1\mathbf{k}]$$
$$q_2 = [s_2, \ x_2\mathbf{i} + y_2\mathbf{j} + z_2\mathbf{k}]$$

Fig. 11.8 The point $P(0, 1, 1)$ is rotated $90°$ about the vector $\mathbf{v}_1$ to $P'(1, 1, 0)$

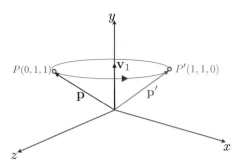

θ is obtained by taking the 4D dot product of q_1 and q_2:

$$\cos\theta = \frac{q_1 \cdot q_2}{|q_1||q_2|}$$
$$= \frac{s_1 s_2 + x_1 x_2 + y_1 y_2 + z_1 z_2}{|q_1||q_2|}$$

and if we are working with unit-norm quaternions, then

$$\cos\theta = s_1 s_2 + x_1 x_2 + y_1 y_2 + z_1 z_2. \tag{11.26}$$

Let's use (11.25) in a scenario with two simple unit-norm quaternions.

Figure 11.8 shows one such scenario where the point $P(0, 1, 1)$ is rotated $90°$ about $\mathbf{v}_1$, the axis of q_1. Figure 11.9 shows another scenario where the same point $P(0, 1, 1)$ is rotated $90°$ about $\mathbf{v}_2$, the axis of q_2. The quaternions are

$$q_1 = \left[\cos 45°, \ \sin 45°\mathbf{j}\right] = \left[\tfrac{\sqrt{2}}{2}, \ \tfrac{\sqrt{2}}{2}\mathbf{j}\right]$$
$$q_2 = \left[\cos 45°, \ \sin 45°\mathbf{i}\right] = \left[\tfrac{\sqrt{2}}{2}, \ \tfrac{\sqrt{2}}{2}\mathbf{i}\right].$$

Therefore, using (11.26)

$$\cos\theta = \tfrac{\sqrt{2}}{2}\tfrac{\sqrt{2}}{2} = 0.5$$
$$\theta = 60°.$$

Before proceeding, let's compute the matrices for the two quaternion products. For q_1:

$$s = \tfrac{\sqrt{2}}{2}, \quad x = 0, \quad y = \tfrac{\sqrt{2}}{2}, \quad z = 0, \quad x_p = 0, \quad y_p = 1, \quad z_p = 1$$

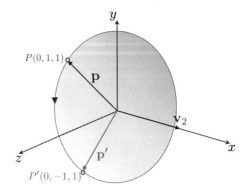

which when substituted in (11.14) gives

$$\mathbf{p}'_1 = \begin{bmatrix} 2(s^2 + x^2) - 1 & 2(xy - sz) & 2(xz + sy) \\ 2(xy + sz) & 2(s^2 + y^2) - 1 & 2(yz - sx) \\ 2(xz - sy) & 2(yz + sx) & 2(s^2 + z^2) - 1 \end{bmatrix} \begin{bmatrix} x_p \\ y_p \\ z_p \end{bmatrix}$$

$$\begin{bmatrix} 1 \\ 1 \\ 0 \end{bmatrix} = \begin{bmatrix} 0 & 0 & 1 \\ 0 & 1 & 0 \\ -1 & 0 & 0 \end{bmatrix} \begin{bmatrix} 0 \\ 1 \\ 1 \end{bmatrix}. \tag{11.27}$$

which is correct.

For q_2:

$$s = \tfrac{\sqrt{2}}{2}, \quad x = \tfrac{\sqrt{2}}{2}, \quad y = 0, \quad z = 0, \quad x_p = 0, \quad y_p = 1, \quad z_p = 1,$$

which when substituted in (11.14) gives

$$\mathbf{p}'_2 = \begin{bmatrix} 2(s^2 + x^2) - 1 & 2(xy - sz) & 2(xz + sy) \\ 2(xy + sz) & 2(s^2 + y^2) - 1 & 2(yz - sx) \\ 2(xz - sy) & 2(yz + sx) & 2(s^2 + z^2) - 1 \end{bmatrix} \begin{bmatrix} x_p \\ y_p \\ z_p \end{bmatrix}$$

$$\begin{bmatrix} 0 \\ -1 \\ 1 \end{bmatrix} = \begin{bmatrix} 1 & 0 & 0 \\ 0 & 0 & -1 \\ 0 & 1 & 0 \end{bmatrix} \begin{bmatrix} 0 \\ 1 \\ 1 \end{bmatrix}. \tag{11.28}$$

which is also correct.

Using (11.25) with $t = 0.5$, computes a mid-way position for an interpolated quaternion, with its vector at 45° between the x- and y-axes, as shown in Fig. 11.10. We already know that $\theta = 60°$, therefore $\sin \theta = \sqrt{3}/2$:

Fig. 11.10 The point $P(0, 1, 1)$ is rotated about the vector **v** to $P'(1, 0, 1)$

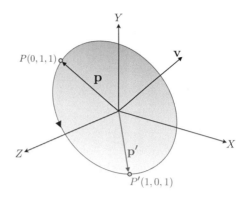

$$q = \frac{\sin[(1-t)\theta]}{\sin\theta}q_1 + \frac{\sin(t\theta)}{\sin\theta}q_2$$

$$= \frac{\sin\left(\frac{60°}{2}\right)}{\sin 60°}\left[\frac{\sqrt{2}}{2}, \frac{\sqrt{2}}{2}\mathbf{j}\right] + \frac{\sin\left(\frac{60°}{2}\right)}{\sin 60°}\left[\frac{\sqrt{2}}{2}, \frac{\sqrt{2}}{2}\mathbf{i}\right]$$

$$= \frac{1}{\sqrt{3}}\left[\frac{\sqrt{2}}{2}, \frac{\sqrt{2}}{2}\mathbf{j}\right] + \frac{1}{\sqrt{3}}\left[\frac{\sqrt{2}}{2}, \frac{\sqrt{2}}{2}\mathbf{i}\right]$$

$$= \left[\frac{\sqrt{2}}{\sqrt{3}}, \frac{1}{\sqrt{6}}\mathbf{i} + \frac{1}{\sqrt{6}}\mathbf{j}\right]$$

where

$$s = \frac{\sqrt{2}}{\sqrt{3}}, \quad x = \frac{1}{\sqrt{6}}, \quad y = \frac{1}{\sqrt{6}}, \quad z = 0, \quad x_p = 0, \quad y_p = 1, \quad z_p = 1,$$

which when substituted in (11.14) gives

$$\mathbf{p'} = \begin{bmatrix} 2(s^2 + x^2) - 1 & 2(xy - sz) & 2(xz + sy) \\ 2(xy + sz) & 2(s^2 + y^2) - 1 & 2(yz - sx) \\ 2(xz - sy) & 2(yz + sx) & 2(s^2 + z^2) - 1 \end{bmatrix} \begin{bmatrix} x_p \\ y_p \\ z_p \end{bmatrix}$$

$$\begin{bmatrix} 1 \\ 0 \\ 1 \end{bmatrix} = \begin{bmatrix} \frac{2}{3} & \frac{1}{3} & \frac{2}{3} \\ \frac{1}{3} & \frac{2}{3} & -\frac{2}{3} \\ -\frac{2}{3} & \frac{2}{3} & \frac{1}{3} \end{bmatrix} \begin{bmatrix} 0 \\ 1 \\ 1 \end{bmatrix} \tag{11.29}$$

which gives the point $P'(1, 0, 1)$.

One of the reasons for using a spherical interpolant is that it linearly interpolates the angle between the two unit-norm quaternions, which creates a constant-angular velocity between them. However, one of the problems with visualising quaternions is that they reside in a four-dimensional space and create a hyper-sphere with a radius

Fig. 11.11 Spherical
interpolation between q_1 and
q_2

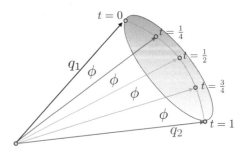

Fig. 11.12 Sketch showing
the actions of the
interpolated quaternions

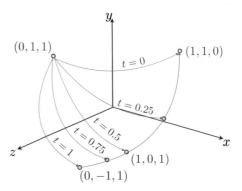

equal to the quaternion's norm. With our 3D brains, this is difficult to visualise.
Nevertheless, we can convince ourselves into thinking we see what is going on with
a simple sketch, as shown in Fig. 11.11, where we see part of the hyper-sphere and
two quaternions q_1 and q_2. In this example, the angle ϕ is a constant angle between
two values of the interpolant t. The spherical interpolant also ensures that the norm
of the interpolated quaternion remains constant at unity, preventing any unwanted
scaling.

Figure 11.12 provides another sketch to help visualise what is going on. For example, when $t = 0$, the interpolated quaternion is q_1 which rotates the point $(0, 1, 1)$ to
$(1, 1, 0)$, and when $t = 1$, the interpolated quaternion is q_2 which rotates the point
$(0, 1, 1)$ to $(0, -1, 1)$. When $t = 0.5$, the interpolated quaternion rotates the point
$(0, 1, 1)$ to $(1, 0, 1)$ as computed above. Two other curves show what happens for
$t = 0.25$ and $t = 0.75$.

A natural consequence of the interpolant is that the angle of rotation is 90° for
$t = 0$ and $t = 1$, but for $t = 0.5$ the angle of rotation (eigenvalue) is approximately
70.5°. Corresponding angles arise for other values of t.

11.9 Converting a Rotation Matrix to a Quaternion

The matrix transform equivalent to qpq^{-1} is

$$
qpq^{-1} = \begin{bmatrix} 2(s^2 + x^2) - 1 & 2(xy - sz) & 2(xz + sy) \\ 2(xy + sz) & 2(s^2 + y^2) - 1 & 2(yz - sx) \\ 2(xz - sy) & 2(yz + sx) & 2(s^2 + z^2) - 1 \end{bmatrix} \begin{bmatrix} x_p \\ y_p \\ z_p \end{bmatrix}
$$

$$
= \begin{bmatrix} a_{11} & a_{12} & a_{13} \\ a_{21} & a_{22} & a_{23} \\ a_{31} & a_{32} & a_{33} \end{bmatrix} \begin{bmatrix} x_p \\ y_p \\ z_p \end{bmatrix}.
$$

Inspection of the matrix shows that by combining various elements we can isolate the terms of a quaternion s, x, y, z. For example, by adding the terms $a_{11} + a_{22} + a_{33}$ we obtain

$$
\begin{aligned}
a_{11} + a_{22} + a_{33} &= \left[2(s^2 + x^2) - 1\right] + \left[2(s^2 + y^2) - 1\right] + \left[2(s^2 + z^2) - 1\right] \\
&= 6s^2 + 2(x^2 + y^2 + z^2) - 3 \\
&= 4s^2 - 1
\end{aligned}
$$

therefore,

$$
s = \pm\tfrac{1}{2}\sqrt{1 + a_{11} + a_{22} + a_{33}}.
$$

To isolate x, y and z we employ

$$
x = \frac{1}{4s}(a_{32} - a_{23})
$$

$$
y = \frac{1}{4s}(a_{13} - a_{31})
$$

$$
z = \frac{1}{4s}(a_{21} - a_{12}).
$$

We can test their correctness using the matrix (11.29):

$$
\begin{bmatrix} a_{11} & a_{12} & a_{13} \\ a_{21} & a_{22} & a_{23} \\ a_{31} & a_{32} & a_{33} \end{bmatrix} = \begin{bmatrix} \frac{2}{3} & \frac{1}{3} & \frac{2}{3} \\ \frac{1}{3} & \frac{2}{3} & -\frac{2}{3} \\ -\frac{2}{3} & \frac{2}{3} & \frac{1}{3} \end{bmatrix}
$$

$$
s = \pm\tfrac{1}{2}\sqrt{1 + a_{11} + a_{22} + a_{33}} = \pm\tfrac{1}{2}\sqrt{1 + \tfrac{2}{3} + \tfrac{2}{3} + \tfrac{1}{3}} = \frac{\sqrt{2}}{\sqrt{3}}
$$

$$
x = \tfrac{1}{4s}(a_{32} - a_{23}) \qquad = \frac{\sqrt{3}}{4\sqrt{2}}\left(\tfrac{2}{3} + \tfrac{2}{3}\right) \qquad = \frac{1}{\sqrt{6}}
$$

$$y = \tfrac{1}{4s}\left(a_{13} - a_{31}\right) \qquad\qquad = \tfrac{\sqrt{3}}{4\sqrt{2}}\left(\tfrac{2}{3} + \tfrac{2}{3}\right) \qquad = \tfrac{1}{\sqrt{6}}$$

$$z = \tfrac{1}{4s}\left(a_{21} - a_{12}\right) \qquad\qquad = \tfrac{\sqrt{3}}{4\sqrt{2}}\left(\tfrac{1}{3} - \tfrac{1}{3}\right) \qquad = 0$$

which agree with the original values.

Say, for example, the value of s had been close to zero, this could have made the values of x, y, z unreliable. Consequently, other combinations are available:

$$x = \pm\tfrac{1}{2}\sqrt{1 + a_{11} - a_{22} - a_{33}}$$

$$y = \frac{1}{4x}\left(a_{12} + a_{21}\right)$$

$$z = \frac{1}{4x}\left(a_{13} + a_{31}\right)$$

$$s = \frac{1}{4x}\left(a_{32} - a_{23}\right).$$

$$y = \pm\tfrac{1}{2}\sqrt{1 - a_{11} + a_{22} - a_{33}}$$

$$x = \frac{1}{4y}\left(a_{12} + a_{21}\right)$$

$$z = \frac{1}{4y}\left(a_{23} + a_{32}\right)$$

$$s = \frac{1}{4y}\left(a_{13} - a_{31}\right).$$

$$z = \pm\tfrac{1}{2}\sqrt{1 - a_{11} - a_{22} + a_{33}}$$

$$x = \frac{1}{4z}\left(a_{13} + a_{31}\right)$$

$$y = \frac{1}{4z}\left(a_{23} + a_{32}\right)$$

$$s = \frac{1}{4z}\left(a_{21} - a_{12}\right).$$

11.10   Euler Angles to Quaternion

Euler angles are rotation transforms about the Cartesian axes: $\mathbf{R}_{\alpha,x}$, $\mathbf{R}_{\beta,y}$ and $\mathbf{R}_{\gamma,z}$, and can be combined to create twelve triple combinations to represent a composite rotation. Let's see how such a transform is represented by a quaternion.

To demonstrate the technique we must choose one of the twelve combinations, then the same technique can be used to convert other combinations. For example, let's choose the sequence $\mathbf{R}_{\gamma,z}\mathbf{R}_{\beta,y}\mathbf{R}_{\alpha,x}$ where the equivalent quaternions are

$$q_x = \left[\cos\left(\tfrac{\alpha}{2}\right),\ \sin\left(\tfrac{\alpha}{2}\right)\mathbf{i}\right]$$
$$q_y = \left[\cos\left(\tfrac{\beta}{2}\right),\ \sin\left(\tfrac{\beta}{2}\right)\mathbf{j}\right]$$
$$q_z = \left[\cos\left(\tfrac{\gamma}{2}\right),\ \sin\left(\tfrac{\gamma}{2}\right)\mathbf{k}\right]$$

and

$$q = q_z q_y q_x. \tag{11.30}$$

Expanding (11.30):

$$q = \left[\cos\left(\tfrac{\gamma}{2}\right),\ \sin\left(\tfrac{\gamma}{2}\right)\mathbf{k}\right]\left[\cos\left(\tfrac{\beta}{2}\right),\ \sin\left(\tfrac{\beta}{2}\right)\mathbf{j}\right]\left[\cos\left(\tfrac{\alpha}{2}\right),\ \sin\left(\tfrac{\alpha}{2}\right)\mathbf{i}\right]$$

$$= \left[\cos\left(\tfrac{\gamma}{2}\right)\cos\left(\tfrac{\beta}{2}\right),\right.$$
$$\left.\cos\left(\tfrac{\gamma}{2}\right)\sin\left(\tfrac{\beta}{2}\right)\mathbf{j} + \cos\left(\tfrac{\beta}{2}\right)\sin\left(\tfrac{\gamma}{2}\right)\mathbf{k} - \sin\left(\tfrac{\gamma}{2}\right)\sin\left(\tfrac{\beta}{2}\right)\mathbf{i}\right]\left[\cos\left(\tfrac{\alpha}{2}\right),\ \sin\left(\tfrac{\alpha}{2}\right)\mathbf{i}\right]$$

$$= \left[\cos\left(\tfrac{\gamma}{2}\right)\cos\left(\tfrac{\beta}{2}\right)\cos\left(\tfrac{\alpha}{2}\right) + \sin\left(\tfrac{\gamma}{2}\right)\sin\left(\tfrac{\beta}{2}\right)\sin\left(\tfrac{\alpha}{2}\right),\right.$$
$$\cos\left(\tfrac{\gamma}{2}\right)\cos\left(\tfrac{\beta}{2}\right)\sin\left(\tfrac{\alpha}{2}\right)\mathbf{i} + \cos\left(\tfrac{\alpha}{2}\right)\cos\left(\tfrac{\gamma}{2}\right)\sin\left(\tfrac{\beta}{2}\right)\mathbf{j} + \cos\left(\tfrac{\alpha}{2}\right)\cos\left(\tfrac{\beta}{2}\right)\sin\left(\tfrac{\gamma}{2}\right)\mathbf{k}$$
$$\left. - \cos\left(\tfrac{\alpha}{2}\right)\sin\left(\tfrac{\gamma}{2}\right)\sin\left(\tfrac{\beta}{2}\right)\mathbf{i} - \cos\left(\tfrac{\gamma}{2}\right)\sin\left(\tfrac{\beta}{2}\right)\sin\left(\tfrac{\alpha}{2}\right)\mathbf{k} + \cos\left(\tfrac{\beta}{2}\right)\sin\left(\tfrac{\gamma}{2}\right)\sin\left(\tfrac{\alpha}{2}\right)\mathbf{j}\right]$$

$$= \left[\cos\left(\tfrac{\gamma}{2}\right)\cos\left(\tfrac{\beta}{2}\right)\cos\left(\tfrac{\alpha}{2}\right) + \sin\left(\tfrac{\gamma}{2}\right)\sin\left(\tfrac{\beta}{2}\right)\sin\left(\tfrac{\alpha}{2}\right),\right.$$
$$\left(\cos\left(\tfrac{\gamma}{2}\right)\cos\left(\tfrac{\beta}{2}\right)\sin\left(\tfrac{\alpha}{2}\right) - \cos\left(\tfrac{\alpha}{2}\right)\sin\left(\tfrac{\gamma}{2}\right)\sin\left(\tfrac{\beta}{2}\right)\right)\mathbf{i}$$
$$+ \left(\cos\left(\tfrac{\alpha}{2}\right)\cos\left(\tfrac{\gamma}{2}\right)\sin\left(\tfrac{\beta}{2}\right) + \cos\left(\tfrac{\beta}{2}\right)\sin\left(\tfrac{\gamma}{2}\right)\sin\left(\tfrac{\alpha}{2}\right)\right)\mathbf{j}$$
$$\left.+ \left(\cos\left(\tfrac{\alpha}{2}\right)\cos\left(\tfrac{\beta}{2}\right)\sin\left(\tfrac{\gamma}{2}\right) - \cos\left(\tfrac{\gamma}{2}\right)\sin\left(\tfrac{\beta}{2}\right)\sin\left(\tfrac{\alpha}{2}\right)\right)\mathbf{k}\right].$$

Now let's place the angles in a consistent sequence:

$$s = \cos\left(\tfrac{\gamma}{2}\right)\cos\left(\tfrac{\beta}{2}\right)\cos\left(\tfrac{\alpha}{2}\right) + \sin\left(\tfrac{\gamma}{2}\right)\sin\left(\tfrac{\beta}{2}\right)\sin\left(\tfrac{\alpha}{2}\right)$$
$$x_q = \cos\left(\tfrac{\gamma}{2}\right)\cos\left(\tfrac{\beta}{2}\right)\sin\left(\tfrac{\alpha}{2}\right) - \sin\left(\tfrac{\gamma}{2}\right)\sin\left(\tfrac{\beta}{2}\right)\cos\left(\tfrac{\alpha}{2}\right)$$
$$y_q = \cos\left(\tfrac{\gamma}{2}\right)\sin\left(\tfrac{\beta}{2}\right)\cos\left(\tfrac{\alpha}{2}\right) + \sin\left(\tfrac{\gamma}{2}\right)\cos\left(\tfrac{\beta}{2}\right)\sin\left(\tfrac{\alpha}{2}\right)$$
$$z_q = \sin\left(\tfrac{\gamma}{2}\right)\cos\left(\tfrac{\beta}{2}\right)\cos\left(\tfrac{\alpha}{2}\right) - \cos\left(\tfrac{\gamma}{2}\right)\sin\left(\tfrac{\beta}{2}\right)\sin\left(\tfrac{\alpha}{2}\right)$$

where

$$q = \left[s,\ x_q\mathbf{i} + y_q\mathbf{j} + z_q\mathbf{k}\right]. \tag{11.31}$$

Let's test (11.31). We start with the three rotation transforms

$$\mathbf{R}_{\alpha,x} = \begin{bmatrix} 1 & 0 & 0 \\ 0 & \cos\alpha & -\sin\alpha \\ 0 & \sin\alpha & \cos\alpha \end{bmatrix}$$

$$\mathbf{R}_{\beta,y} = \begin{bmatrix} \cos\beta & 0 & \sin\beta \\ 0 & 1 & 0 \\ -\sin\beta & 0 & \cos\beta \end{bmatrix}$$

$$\mathbf{R}_{\gamma,z} = \begin{bmatrix} \cos\gamma & -\sin\gamma & 0 \\ \sin\gamma & \cos\gamma & 0 \\ 0 & 0 & 1 \end{bmatrix}.$$

Then

$$\mathbf{R}_{\gamma,z}\mathbf{R}_{\beta,y}\mathbf{R}_{\alpha,x} =$$
$$\begin{bmatrix} \cos\gamma\cos\beta & -\sin\gamma\cos\alpha + \cos\gamma\sin\beta\sin\alpha & \sin\gamma\sin\alpha + \cos\gamma\sin\beta\cos\alpha \\ \sin\gamma\cos\beta & \cos\gamma\cos\alpha + \sin\gamma\sin\beta\sin\alpha & -\cos\gamma\sin\alpha + \sin\gamma\sin\beta\cos\alpha \\ -\sin\beta & \cos\beta\sin\alpha & \cos\beta\cos\alpha \end{bmatrix}.$$

Let's make $\alpha = \beta = \gamma = 90°$, then

$$\mathbf{R}_{90°,z}\mathbf{R}_{90°,y}\mathbf{R}_{90°,x} = \begin{bmatrix} 0 & 0 & 1 \\ 0 & 1 & 0 \\ -1 & 0 & 0 \end{bmatrix}$$

which rotates points 90° about the y-axis:

$$\begin{bmatrix} 1 \\ 1 \\ 0 \end{bmatrix} = \begin{bmatrix} 0 & 0 & 1 \\ 0 & 1 & 0 \\ -1 & 0 & 0 \end{bmatrix}\begin{bmatrix} 0 \\ 1 \\ 1 \end{bmatrix}.$$

Now let's evaluate (11.31):

$$s = \cos\left(\tfrac{\gamma}{2}\right)\cos\left(\tfrac{\beta}{2}\right)\cos\left(\tfrac{\alpha}{2}\right) + \sin\left(\tfrac{\gamma}{2}\right)\sin\left(\tfrac{\beta}{2}\right)\sin\left(\tfrac{\alpha}{2}\right)$$
$$= \tfrac{\sqrt{2}}{2}\tfrac{\sqrt{2}}{2}\tfrac{\sqrt{2}}{2} + \tfrac{\sqrt{2}}{2}\tfrac{\sqrt{2}}{2}\tfrac{\sqrt{2}}{2}$$
$$= \tfrac{\sqrt{2}}{2}$$
$$x_q = \cos\left(\tfrac{\gamma}{2}\right)\cos\left(\tfrac{\beta}{2}\right)\sin\left(\tfrac{\alpha}{2}\right) - \sin\left(\tfrac{\gamma}{2}\right)\sin\left(\tfrac{\beta}{2}\right)\cos\left(\tfrac{\alpha}{2}\right)$$
$$= \tfrac{\sqrt{2}}{2}\tfrac{\sqrt{2}}{2}\tfrac{\sqrt{2}}{2} - \tfrac{\sqrt{2}}{2}\tfrac{\sqrt{2}}{2}\tfrac{\sqrt{2}}{2}$$
$$= 0$$

$$y_q = \cos\left(\tfrac{\gamma}{2}\right)\sin\left(\tfrac{\beta}{2}\right)\cos\left(\tfrac{\alpha}{2}\right) + \sin\left(\tfrac{\gamma}{2}\right)\cos\left(\tfrac{\beta}{2}\right)\sin\left(\tfrac{\alpha}{2}\right)$$

$$= \tfrac{\sqrt{2}}{2}\tfrac{\sqrt{2}}{2}\tfrac{\sqrt{2}}{2} + \tfrac{\sqrt{2}}{2}\tfrac{\sqrt{2}}{2}\tfrac{\sqrt{2}}{2}$$

$$= \tfrac{\sqrt{2}}{2}$$

$$z_q = \sin\left(\tfrac{\gamma}{2}\right)\cos\left(\tfrac{\beta}{2}\right)\cos\left(\tfrac{\alpha}{2}\right) - \cos\left(\tfrac{\gamma}{2}\right)\sin\left(\tfrac{\beta}{2}\right)\sin\left(\tfrac{\alpha}{2}\right)$$

$$= \tfrac{\sqrt{2}}{2}\tfrac{\sqrt{2}}{2}\tfrac{\sqrt{2}}{2} - \tfrac{\sqrt{2}}{2}\tfrac{\sqrt{2}}{2}\tfrac{\sqrt{2}}{2}$$

$$= 0$$

and

$$q = \left[\tfrac{\sqrt{2}}{2},\ \tfrac{\sqrt{2}}{2}\mathbf{j}\right]$$

which is a quaternion that also rotates points 90° about the y-axis.

11.11 Summary

In this chapter I have shown how unit-norm quaternions can be used to rotate a vector about a quaternion's vector. It would have been useful if this could have been achieved by the simple product qp, like complex numbers. But as we saw, this only works when the quaternion is orthogonal to the vector. The product qpq^{-1} – discovered by Hamilton and Cayley – works for all orientations between a quaternion and a vector. We also saw that the product can be represented as a matrix, which can be integrated with other matrices, and implemented in software.

Perhaps one of the most interesting features of quaternions that has emerged in this chapter, is that their imaginary qualities are not required in any calculations, because they are embedded within the algebra.

The spherical interpolant provides a clever way to dynamically change a quaternion's axis and angle of rotation, but can be difficult to visualise as an animated sequence without access to a real-time display system.

The reverse product $q^{-1}pq$ reverses the angle of rotation, and is equivalent to changing the sign of the rotation angle in qpq^{-1}. Consequently, it can be used to rotate a frame of reference in the same direction as qpq^{-1}.

11.11.1 Summary of Operations

Rotating a Point About a Vector

$$q = [s,\ \mathbf{v}]$$
$$s^2 + |\mathbf{v}|^2 = 1$$

$$p = [0, \ \mathbf{p}]$$
$$qpq^{-1} = \left[0, \ 2(\mathbf{v} \cdot \mathbf{p})\mathbf{v} + \left(2s^2 - 1\right)\mathbf{p} + 2s\mathbf{v} \times \mathbf{p}\right]$$
$$q = \left[\cos\left(\tfrac{\theta}{2}\right), \ \sin\left(\tfrac{\theta}{2}\right)\hat{\mathbf{v}}\right]$$
$$p = [0, \ \mathbf{p}]$$
$$qpq^{-1} = \left[0, \ (1 - \cos\theta)(\hat{\mathbf{v}} \cdot \mathbf{p})\hat{\mathbf{v}} + \cos\theta\mathbf{p} + \sin\theta\hat{\mathbf{v}} \times \mathbf{p}\right].$$

Matrix for Rotating a Point about a Vector

$$\mathbf{p}' = \begin{bmatrix} 1 - 2\left(y^2 + z^2\right) & 2(xy - sz) & 2(xz + sy) \\ 2(xy + sz) & 1 - 2\left(x^2 + z^2\right) & 2(yz - sx) \\ 2(xz - sy) & 2(yz + sx) & 1 - 2\left(x^2 + y^2\right) \end{bmatrix} \begin{bmatrix} x_p \\ y_p \\ z_p \end{bmatrix}.$$

Matrix for a Quaternion Product

$$q_1 q_2 = \mathbf{L}(q_1)q_2 = \begin{bmatrix} s_1 & -x_1 & -y_1 & -z_1 \\ x_1 & s_1 & -z_1 & y_1 \\ y_1 & z_1 & s_1 & -x_1 \\ z_1 & -y_1 & x_1 & s_1 \end{bmatrix} \begin{bmatrix} s_2 \\ x_2 \\ y_2 \\ z_2 \end{bmatrix}$$

$$q_1 q_2 = \mathbf{R}(q_2)q_1 = \begin{bmatrix} s_2 & -x_2 & -y_2 & -z_2 \\ x_2 & s_2 & z_2 & -y_2 \\ y_2 & -z_2 & s_2 & x_2 \\ z_2 & y_2 & -x_2 & s_2 \end{bmatrix} \begin{bmatrix} s_1 \\ x_1 \\ y_1 \\ z_1 \end{bmatrix}.$$

Interpolating Two Quaternions

$$q = \frac{\sin[(1 - t)\theta]}{\sin\theta}q_1 + \frac{\sin(t\theta)}{\sin\theta}q_2$$

where

$$\cos\theta = \frac{q_1 \cdot q_2}{|q_1||q_2|}$$
$$= \frac{s_1 s_2 + x_1 x_2 + y_1 y_2 + z_1 z_2}{|q_1||q_2|}.$$

Quaternion from a Rotation Matrix

$$s = \pm\frac{1}{2}\sqrt{1 + a_{11} + a_{22} + a_{33}}$$

$$x = \frac{1}{4s}\left(a_{32} - a_{23}\right)$$

$$y = \frac{1}{4s}\left(a_{13} - a_{31}\right)$$

$$z = \frac{1}{4s}\left(a_{21} - a_{12}\right).$$

$$x = \pm\frac{1}{2}\sqrt{1 + a_{11} - a_{22} - a_{33}}$$

$$y = \frac{1}{4x}\left(a_{12} + a_{21}\right)$$

$$z = \frac{1}{4x}\left(a_{13} + a_{31}\right)$$

$$s = \frac{1}{4x}\left(a_{32} - a_{23}\right).$$

$$y = \pm\frac{1}{2}\sqrt{1 - a_{11} + a_{22} - a_{33}}$$

$$x = \frac{1}{4y}\left(a_{12} + a_{21}\right)$$

$$z = \frac{1}{4y}\left(a_{23} + a_{32}\right)$$

$$s = \frac{1}{4y}\left(a_{13} - a_{31}\right).$$

$$z = \pm\frac{1}{2}\sqrt{1 - a_{11} - a_{22} + a_{33}}$$

$$x = \frac{1}{4z}\left(a_{13} + a_{31}\right)$$

$$y = \frac{1}{4z}\left(a_{23} + a_{32}\right)$$

$$s = \frac{1}{4z}\left(a_{21} - a_{12}\right).$$

Eigenvector and Eigenvalue

$$x_v = \left(a_{22} - 1\right)\left(a_{33} - 1\right) - a_{23}a_{32}$$

$$y_v = \left(a_{33} - 1\right)\left(a_{11} - 1\right) - a_{31}a_{13}$$

$$z_v = \left(a_{11} - 1\right)\left(a_{22} - 1\right) - a_{12}a_{21}$$

$$\cos\theta = \tfrac{1}{2}\left(\mathrm{Tr}\left(qpq^{-1}\right) - 1\right).$$

Euler Angles to Quaternion
Using the transform $\mathbf{R}_{\gamma,z}\mathbf{R}_{\beta,y}\mathbf{R}_{\alpha,x}$:

$$s = \cos\left(\tfrac{\gamma}{2}\right)\cos\left(\tfrac{\beta}{2}\right)\cos\left(\tfrac{\alpha}{2}\right) + \sin\left(\tfrac{\gamma}{2}\right)\sin\left(\tfrac{\beta}{2}\right)\sin\left(\tfrac{\alpha}{2}\right)$$

$$x_q = \cos\left(\tfrac{\gamma}{2}\right)\cos\left(\tfrac{\beta}{2}\right)\sin\left(\tfrac{\alpha}{2}\right) - \sin\left(\tfrac{\gamma}{2}\right)\sin\left(\tfrac{\beta}{2}\right)\cos\left(\tfrac{\alpha}{2}\right)$$

$$y_q = \cos\left(\tfrac{\gamma}{2}\right)\sin\left(\tfrac{\beta}{2}\right)\cos\left(\tfrac{\alpha}{2}\right) + \sin\left(\tfrac{\gamma}{2}\right)\cos\left(\tfrac{\beta}{2}\right)\sin\left(\tfrac{\alpha}{2}\right)$$

$$z_q = \sin\left(\tfrac{\gamma}{2}\right)\cos\left(\tfrac{\beta}{2}\right)\cos\left(\tfrac{\alpha}{2}\right) - \cos\left(\tfrac{\gamma}{2}\right)\sin\left(\tfrac{\beta}{2}\right)\sin\left(\tfrac{\alpha}{2}\right)$$

where

$$q = \left[s,\ x_q\mathbf{i} + y_q\mathbf{j} + z_q\mathbf{k}\right].$$

11.12 Worked Examples

11.12.1 Rotate a Vector Using qp

Use the product qp to rotate $p = [0,\ \mathbf{j}]$, $90°$ about the x-axis.
   Solution: For this to work, q must be orthogonal to p:

$$q = [\cos\theta,\ \sin\theta\mathbf{i}]$$
$$= [0,\ \mathbf{i}]$$

and

$$p' = qp$$
$$= [0,\ \mathbf{i}][0,\ \mathbf{j}]$$
$$= [0,\ \mathbf{k}].$$

11.12.2 Rotate a Vector Using qpq^{-1}

Use the product qpq^{-1} to rotate $p = [0,\ \mathbf{j}]$, $90°$ about the x-axis.

Solution: For this to work:

$$q = \left[\cos\left(\tfrac{\theta}{2}\right),\ \sin\left(\tfrac{\theta}{2}\right)\mathbf{i}\right]$$
$$= \left[\tfrac{\sqrt{2}}{2},\ \tfrac{\sqrt{2}}{2}\mathbf{i}\right]$$

and

$$p' = qpq^{-1}$$
$$= \left[\tfrac{\sqrt{2}}{2},\ \tfrac{\sqrt{2}}{2}\mathbf{i}\right][0,\ \mathbf{j}]\left[\tfrac{\sqrt{2}}{2},\ -\tfrac{\sqrt{2}}{2}\mathbf{i}\right]$$
$$= \left[0,\ \tfrac{\sqrt{2}}{2}\mathbf{j} + \tfrac{\sqrt{2}}{2}\mathbf{k}\right]\left[\tfrac{\sqrt{2}}{2},\ -\tfrac{\sqrt{2}}{2}\mathbf{i}\right]$$
$$= \left[0,\ \tfrac{\sqrt{2}}{2}\left(\tfrac{\sqrt{2}}{2}\mathbf{j} + \tfrac{\sqrt{2}}{2}\mathbf{k}\right) + \tfrac{1}{2}\mathbf{j} + \tfrac{1}{2}\mathbf{k}\right]$$
$$= \left[0,\ \tfrac{1}{2}\mathbf{j} + \tfrac{1}{2}\mathbf{k} - \tfrac{1}{2}\mathbf{j} + \tfrac{1}{2}\mathbf{k}\right]$$
$$= [0,\ \mathbf{k}]$$

which is correct.

11.12.3   Rotate a Vector 360° Using qpq^{-1}

Evaluate the product qpq^{-1} for $p = [0,\ \mathbf{p}]$ and $q = \left[\cos\left(\tfrac{\theta}{2}\right),\ \sin\left(\tfrac{\theta}{2}\right)\mathbf{v}\right]$, where $\theta = 360°$.
   Solution:

$$q = [-1,\ \mathbf{0}]$$
$$qpq^{-1} = [-1,\ \mathbf{0}][0,\ \mathbf{p}][-1,\ \mathbf{0}]$$
$$= [0,\ -\mathbf{p}][-1,\ \mathbf{0}]$$
$$= [0,\ \mathbf{p}]$$

which confirms that the vector remains unmoved, as expected.

11.12.4   Quaternion as a Matrix

Compute the matrix (11.14) for $q = \left[\tfrac{1}{2},\ \tfrac{\sqrt{3}}{2}\mathbf{k}\right]$, and find its eigenvector and eigenvalue.

Solution: From q:

$$s = \tfrac{1}{2}, \quad x = 0, \quad y = 0, \quad z = \tfrac{\sqrt{3}}{2}$$

$$
\mathbf{p}' = \begin{bmatrix} 2(s^2 + x^2) - 1 & 2(xy - sz) & 2(xz + sy) \\ 2(xy + sz) & 2(s^2 + y^2) - 1 & 2(yz - sx) \\ 2(xz - sy) & 2(yz + sx) & 2(s^2 + z^2) - 1 \end{bmatrix} \begin{bmatrix} x_p \\ y_p \\ z_p \end{bmatrix}
$$

$$
= \begin{bmatrix} -\tfrac{1}{2} & -\tfrac{\sqrt{3}}{2} & 0 \\ \tfrac{\sqrt{3}}{2} & -\tfrac{1}{2} & 0 \\ 0 & 0 & 1 \end{bmatrix} \begin{bmatrix} x_p \\ y_p \\ z_p \end{bmatrix}.
$$

If we plug in the point $(1, 0, 0)$ it is rotated about the z-axis by $120°$:

$$
\begin{bmatrix} -\tfrac{1}{2} \\ \tfrac{\sqrt{3}}{2} \\ 1 \end{bmatrix} = \begin{bmatrix} -\tfrac{1}{2} & -\tfrac{\sqrt{3}}{2} & 0 \\ \tfrac{\sqrt{3}}{2} & -\tfrac{1}{2} & 0 \\ 0 & 0 & 1 \end{bmatrix} \begin{bmatrix} 1 \\ 0 \\ 0 \end{bmatrix}.
$$

Using

$$
\begin{aligned}
\cos\theta &= \tfrac{1}{2}\left(\mathrm{Tr}(qpq^{-1}) - 1\right) \\
&= \tfrac{1}{2}(0 - 1) \\
\theta &= 120°.
\end{aligned}
$$

Using

$$
\begin{aligned}
x_v &= (a_{22} - 1)(a_{33} - 1) - a_{23}a_{32} \\
&= \left(-\tfrac{3}{2}\right)(0) - 0 \\
&= 0 \\
y_v &= (a_{33} - 1)(a_{11} - 1) - a_{31}a_{13} \\
&= (0)\left(-\tfrac{3}{2}\right) - 0 \\
&= 0 \\
z_v &= (a_{11} - 1)(a_{22} - 1) - a_{12}a_{21} \\
&= \left(-\tfrac{3}{2}\right)\left(-\tfrac{3}{2}\right) + \tfrac{\sqrt{3}}{2}\tfrac{\sqrt{3}}{2} \\
&= 3
\end{aligned}
$$

which makes the eigenvector $3\mathbf{k}$ and the eigenvalue $120°$.

11.12.5 Interpolating a Quaternion

Find the half-way quaternion between $q_1 = \left[\cos\left(\frac{\alpha}{2}\right),\ \sin\left(\frac{\alpha}{2}\right)\mathbf{k}\right]$ and $q_2 = \left[\cos\left(\frac{\alpha}{2}\right),\ \sin\left(\frac{\alpha}{2}\right)\mathbf{i}\right]$ when $\alpha = 90°$. Show that it is a unit-norm quaternion, and find its angle of rotation.

Solution: The angle between q_1 and q_2 is θ where

$$\cos\theta = \frac{s_1 s_2 + x_1 x_2 + y_1 y_2 + z_1 z_2}{|q_1||q_2|}$$
$$= \cos^2 \tfrac{\alpha}{2}$$
$$= 0.5$$
$$\theta = 60°.$$

Using

$$q = \frac{\sin((1-t)\theta)}{\sin\theta} q_1 + \frac{\sin(t\theta)}{\sin\theta} q_2$$
$$= \frac{\sin 30°}{\sin 60°}\left[\cos 45°,\ \sin 45°\mathbf{k}\right] + \frac{\sin 30°}{\sin 60°}\left[\cos 45°,\ \sin 45°\mathbf{i}\right]$$
$$= \frac{1}{\sqrt{3}}\left[\tfrac{\sqrt{2}}{2},\ \tfrac{\sqrt{2}}{2}\mathbf{k}\right] + \frac{1}{\sqrt{3}}\left[\tfrac{\sqrt{2}}{2},\ \tfrac{\sqrt{2}}{2}\mathbf{i}\right]$$
$$= \left[\tfrac{\sqrt{2}}{\sqrt{3}},\ \tfrac{\sqrt{2}}{2\sqrt{3}}\mathbf{i} + \tfrac{\sqrt{2}}{2\sqrt{3}}\mathbf{k}\right]$$
$$= \left[\tfrac{2}{\sqrt{6}},\ \tfrac{1}{\sqrt{6}}\mathbf{i} + \tfrac{1}{\sqrt{6}}\mathbf{k}\right].$$

The norm of q is

$$|q| = \left(\tfrac{2}{\sqrt{6}}\right)^2 + \left(\tfrac{1}{\sqrt{6}}\right)^2 + \left(\tfrac{1}{\sqrt{6}}\right)^2$$
$$= \tfrac{2}{3} + \tfrac{1}{6} + \tfrac{1}{6}$$
$$= 1.$$

Therefore, $\cos\left(\frac{\alpha}{2}\right) = \frac{\sqrt{2}}{\sqrt{3}}$ and $\sin\left(\frac{\alpha}{2}\right) = \frac{1}{\sqrt{3}}$, and $\alpha \approx 70.5°$.

11.12.6 Convert a Rotation Matrix as a Quaternion

Convert the matrix $\mathbf{M}$ into a quaternion and identify its function.

Solution:

$$\mathbf{M} = \begin{bmatrix} 0 & 0 & 1 \\ 0 & 1 & 0 \\ -1 & 0 & 0 \end{bmatrix}$$

therefore,

$$s = \tfrac{1}{2}\sqrt{1 + a_{11} + a_{22} + a_{33}}$$
$$= \tfrac{1}{2}\sqrt{1 + 0 + 1 + 0} \quad = \tfrac{\sqrt{2}}{2}$$
$$x = \frac{1}{4s}\left(a_{32} - a_{23}\right)$$
$$= \tfrac{\sqrt{2}}{4}\left(0 + 0\right) \quad = 0$$
$$y = \frac{1}{4s}\left(a_{13} - a_{31}\right)$$
$$= \tfrac{\sqrt{2}}{4}\left(1 + 1\right) \quad = \tfrac{\sqrt{2}}{2}$$
$$z = \frac{1}{4s}\left(a_{21} - a_{12}\right)$$
$$= \tfrac{\sqrt{2}}{4}\left(0 + 0\right) \quad = 0$$

which is the quaternion $\left[\tfrac{\sqrt{2}}{2},\ \tfrac{\sqrt{2}}{2}\mathbf{j}\right]$, and is a rotation of 90° about the y-axis.

References

1. Rodrigues BO (1840) Des lois géométriques qui régissent les déplacements d'un système solide dans l'espace, et de la variation des coordonnées provent de ses déplacements considérés indépendamment des causes qui peuvent les produire. J. de Matématiques Pures et Appliquées 5:380–440
2. Cayley A (1848) The collected mathematical papers, vol I, p 586, note 20
3. Altmann S L (1986) Rotations, quaternions and double groups. Dover Publications (2005), p. 16, ISBN-13: 978-0-486-44518-2

Chapter 12
Complex Numbers and the Riemann Hypothesis

12.1 Introduction

Prime numbers are very easy to define: a number whose factors are 1 and itself; the first nine primes being 2, 3, 5, 7, 11, 13, 17, 19, 23. However, in spite of this simple definition, it has been impossible to find a formula that predicts primes, or even count the exact number of primes up to some limit. Many mathematicians have taken up the challenge, but all have failed. Nevertheless, their endeavours have been astounding and created some incredible results, formulae and conjectures. This chapter outlines the work of Leonhard Euler, and the brilliant German mathematician Bernhard Riemann (1826–1866), and his famous hypothesis.

12.2 Euler's Work

12.2.1 Euler's Zeta Function

Euler was aware that

$$\sin x = \frac{x}{1!} - \frac{x^3}{3!} + \frac{x^5}{5!} - \frac{x^7}{7!} + \cdots$$

and therefore,

$$\frac{\sin x}{x} = 1 - \frac{x^2}{3!} + \frac{x^4}{5!} - \frac{x^6}{7!} + \cdots. \qquad (12.1)$$

© Springer International Publishing AG, part of Springer Nature 2018
J. Vince, *Imaginary Mathematics for Computer Science*,
https://doi.org/10.1007/978-3-319-94637-5_12

Factorising (12.1) using the Weierstrass factorisation theorem [1], we obtain

$$\frac{\sin x}{x} = \left(1 - \frac{x}{\pi}\right)\left(1 + \frac{x}{\pi}\right)\left(1 - \frac{x}{2\pi}\right)\left(1 + \frac{x}{2\pi}\right)\left(1 - \frac{x}{3\pi}\right)\left(1 + \frac{x}{3\pi}\right)\cdots$$

$$= \left(1 - \frac{x^2}{\pi^2}\right)\left(1 - \frac{x^2}{4\pi^2}\right)\left(1 - \frac{x^2}{9\pi^2}\right)\left(1 - \frac{x^2}{16\pi^2}\right)\cdots. \tag{12.2}$$

The x^2 coefficient of (12.1) is $-\frac{1}{6}$, which must equal the x^2 coeffiicent of (12.2):

$$-\frac{1}{6} = -\left(\frac{1}{\pi^2} + \frac{1}{4\pi^2} + \frac{1}{9\pi^2} + \frac{1}{16\pi^2} + \cdots\right)$$

$$= -\frac{1}{\pi^2}\left(\frac{1}{1^2} + \frac{1}{2^2} + \frac{1}{3^2} + \frac{1}{4^2} + \cdots\right)$$

therefore,

$$\frac{\pi^2}{6} = \frac{1}{1^2} + \frac{1}{2^2} + \frac{1}{3^2} + \frac{1}{4^2} + \cdots = \sum_{n=1}^{\infty} \frac{1}{n^2}. \tag{12.3}$$

Equations (12.3) and (12.4) show Euler's zeta function, which was the starting point for Riemann. It is defined for any real number s greater than 1 by the infinite sum:

$$\zeta(s) = 1 + \frac{1}{2^s} + \frac{1}{3^s} + \frac{1}{4^s} + \frac{1}{5^s} + \cdots. \tag{12.4}$$

When $s = 1$ we get the harmonic series:

$$\zeta(1) = 1 + \frac{1}{2} + \frac{1}{3} + \frac{1}{4} + \frac{1}{5} + \cdots$$

which diverges to infinity.

Euler's extraordinary algebraic skills revealed how his zeta function is linked to the primes. For example, he found individual sums for the odd and even terms in (12.3) as follows:

$$\frac{\pi^2}{6} = \frac{1}{1^2} + \frac{1}{2^2} + \frac{1}{3^2} + \frac{1}{4^2} + \cdots \tag{12.5}$$

multiplying both sides by $\frac{1}{2^2}$:

$$\frac{1}{2^2}\frac{\pi^2}{6} = \frac{1}{2^2}\frac{1}{1^2} + \frac{1}{2^2}\frac{1}{2^2} + \frac{1}{2^2}\frac{1}{3^2} + \frac{1}{2^2}\frac{1}{4^2} + \cdots$$

$$= \frac{1}{2^2} + \frac{1}{4^2} + \frac{1}{6^2} + \frac{1}{8^2} + \cdots \qquad (12.6)$$

subtracting (12.6) from (12.5):

$$\left(1 - \frac{1}{2^2}\right)\frac{\pi^2}{6} = \frac{1}{1^2} + \frac{1}{3^2} + \frac{1}{5^2} + \frac{1}{7^2} + \cdots$$

$$\frac{\pi^2}{8} = \frac{1}{1^2} + \frac{1}{3^2} + \frac{1}{5^2} + \frac{1}{7^2} + \cdots \qquad (12.7)$$

and subtracting (12.7) from (12.5):

$$\frac{\pi^2}{24} = \frac{1}{2^2} + \frac{1}{4^2} + \frac{1}{6^2} + \frac{1}{8^2} + \cdots .$$

Employing the same algebraic strategy with the zeta function, we have

$$\zeta(s) = \frac{1}{1^s} + \frac{1}{2^s} + \frac{1}{3^s} + \frac{1}{4^s} + \frac{1}{5^s} + \cdots$$

$$\left(1 - \frac{1}{2^s}\right)\zeta(s) = \frac{1}{1^s} + \frac{1}{3^s} + \frac{1}{5^s} + \frac{1}{7^s} + \frac{1}{9^s} + \cdots \qquad (12.8)$$

$$\left(\frac{1}{2^s}\right)\zeta(s) = \frac{1}{2^s} + \frac{1}{4^s} + \frac{1}{6^s} + \frac{1}{8^s} + \frac{1}{10^s} + \cdots . \qquad (12.9)$$

Euler then used (12.8) and (12.9) to develop his famous prime product formula by multiplying (12.8) by $\frac{1}{3^s}$:

$$\frac{1}{3^s}\left(1 - \frac{1}{2^s}\right)\zeta(s) = \frac{1}{3^s}\frac{1}{1^s} + \frac{1}{3^s}\frac{1}{3^s} + \frac{1}{3^s}\frac{1}{5^s} + \frac{1}{3^s}\frac{1}{7^s} + \cdots$$

$$= \frac{1}{3^s} + \frac{1}{9^s} + \frac{1}{15^s} + \frac{1}{21^s} + \frac{1}{27^s} + \cdots . \qquad (12.10)$$

Subtracting (12.10) from (12.8) we have

$$\left(1 - \frac{1}{3^s}\right)\left(1 - \frac{1}{2^s}\right)\zeta(s) = \frac{1}{1^s} + \frac{1}{5^s} + \frac{1}{7^s} + \frac{1}{11^s} + \cdots \qquad (12.11)$$

By repeating this process for the remaining terms on the RHS of (12.11), in the limit, we have

$$\cdots \left(1 - \frac{1}{7^s}\right)\left(1 - \frac{1}{5^s}\right)\left(1 - \frac{1}{3^s}\right)\left(1 - \frac{1}{2^s}\right)\zeta(s) = 1$$

and

$$\zeta(s) = \frac{1}{\left(1 - \frac{1}{2^s}\right)\left(1 - \frac{1}{3^s}\right)\left(1 - \frac{1}{5^s}\right)\left(1 - \frac{1}{7^s}\right)\cdots}.$$

But

$$\sum_{n}^{\infty} \frac{1}{n^s} = \zeta(s)$$

and

$$\prod_{p} \frac{1}{1 - \frac{1}{p^s}} = \frac{1}{\left(1 - \frac{1}{2^s}\right)\left(1 - \frac{1}{3^s}\right)\left(1 - \frac{1}{5^s}\right)\left(1 - \frac{1}{7^s}\right)\cdots}$$

therefore,

$$\prod_{p} \frac{1}{1 - \frac{1}{p^s}} = \sum_{n}^{\infty} \frac{1}{n^s}$$

which is Euler's prime product formula, and was the starting point for Riemann's seminal 1859 paper *Ueber die Anzahl der Primzahlen unter einer gegebenen Grösse*, which translated is *About the Number of Prime Numbers Under a Given Number* [2].

12.3   The Prime Number Theorem

An approximate formula for counting primes up to a value x, is given by

$$\pi(x) \sim \frac{x}{\ln x}.$$

Table 12.1 shows the percentage error for increasing values of x.

Table 12.1 $\pi(x)$ and x/lnx for different values of x

x	$\pi(x)$	x/lnx	% error
10^2	25	22	-12
10^3	168	145	-13.7
10^6	78,498	72,382	-7.8
10^9	50,847,534	48,254,942	-5.1

Table 12.2 $\pi(x)$ and $\text{Li}(x) - \pi(x)$ for different values of x

x	$\pi(x)$	$\text{Li}(x) - \pi(x)$	% error
10^8	5,761,455	754	0.013
10^9	50,847,534	1, 701	0.0033
10^{10}	455,052,511	3, 104	0.00068
10^{14}	53,204,941,750,802	314,890	0.0000006

Riemann's paper introduced an improved prime number theorem:

$$\pi(x) \sim \text{Li}(x)$$

where

$$\text{Li}(x) = \int_2^x \frac{1}{\ln(t)} dt.$$

Table 12.2 shows the accuracy of the prime number theorem at counting primes. Thus $\pi(x)$ lies between an upper value of $\text{Li}(x)$, and a lower value of $x/\ln x$.

12.4 The Riemann Zeta Function

Euler's zeta function $\zeta(s)$ assumes that $s > 1$, where $s \in \mathbb{R}$, and is convergent. The breakthrough made by Riemann was to make s complex: $s \in \mathbb{C}$. Riemann was particularly interested in when $\zeta(s) = 0$, and showed that Euler's prime product formula converges for complex s, with $\text{Re}(s) > 1$, and therefore has no zeros in the region marked yellow in Fig. 12.1.

The infinite sum form of the zeta function is not very useful when one is looking for particular values of s that make $\zeta(s) = 0$, which is why Riemann proposed a functional form (12.12) in his paper:

$$\zeta(s) = 2^s \pi^{s-1} \sin\left(\frac{\pi s}{2}\right) \Gamma(1-s)\zeta(1-s) \tag{12.12}$$

Fig. 12.1 The complex
plane for the zeta function

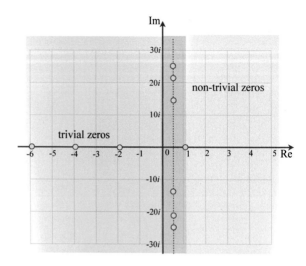

where the gamma function $\Gamma(n) = (n - 1)!$

Equation (12.12) applies to the entire complex plane, and one can see in the green zone of Fig. 12.1 values of s that make $\zeta(s) = 0$; these are called "trivial zeros", and arise when $\sin(\pi s/2) = 0$ in (12.12). i.e. when $s = -2n$. Thus there exists an infinite number of points along the negative real axis, where s makes $\zeta(s) = 0$.

Riemann also showed that other values of s exist on a critical line $\text{Re}(s) = 1/2$, where $\zeta(s) = 0$; these are called "non-trivial zeros". Furthermore, these occur in conjugate pairs. i.e. $1/2 \pm bi$. In the critical red strip of Fig. 12.1, the first such non-trivial zero is $1/2 + 14.134725\ldots i$. Others include $1/2 + 21.022040\ldots i$, $1/2 + 25.010856\ldots i$, $1/2 + 30.425$ and $1/2 + 32.935\ldots i$.

The other point marked on Fig. 12.1 is $s = 1$, which makes $\zeta(1) = \infty$.

12.4.1 *The Riemann Hypothesis*

Having gone through the above explanation, the Riemann hypothesis is extremely simple to state, and is the conjecture that "the zeta function is zero only at the negative even integers (trivial zeros), and complex numbers s with $\text{Re}(s) = 1/2$, (non-trivial zeros)".

Some mathematicians regard the Riemann conjecture as a major unsolved problem, and that its proof will influence other branches of mathematics, especially the distribution of prime numbers. Others believe that it is false. So far, no one has been able to prove or disprove the conjecture. Hopefully, the prize of prize of $\$10^6$, offered by the Clay Mathematics Institute, will keep mathematicians trying.

References

1. https://en.wikipedia.org/wiki/Basel_problem
2. Riemann B (1859) Ueber die Anzahl der Primzahlen unter einer gegebenen Grsse, Monats-
   berichte der Berliner Akademie. In Gesammelte Werke, Teubner, Leipzig (1892), Reprinted by
   Dover, New York (1953). Original manuscript (with English translation). Reprinted in (Borwein
   et al 2008) and (Edwards 1974)

Chapter 13
The Mandelbrot Set

13.1 Introduction

The Mandelbrot set is a simple application of complex numbers that reveals an amazing degree of complexity that would have remained hidden without the digital computer. I remember discovering the set in the mid-1980s, and tried running the program on my BBC micro. It took forever, and was not worth the wait. Today, things are completely different, and many beautiful images are possible in a fraction of a second.

13.2 The Mandelbrot Set

The Polish born, mathematician Benoit Mandelbrot (1924–2010), is recognised for inventing the term *fractal* in 1975, which describe structures possessing self-similar properties. During his time at IBM as an IBM Fellow, Mandelbrot researched the subject of fractal geometry, and a particular image known as the *Mandelbrot Set*, Fig. 13.1 is named after him.

The Mandelbrot set $\mathbb{M}$ is a set of complex numbers that pass a specific test, that determines whether a complex number is inside or outside $\mathbb{M}$. As the Mandelbrot set possesses an infinite level of detail and complexity, the test is subject to two conditions: the first, normally constrains the test to a radius of 2 on the complex plane; the second, controls the number of iterations made by the algorithm to conduct the test. The points on the complex plane are assigned colours according to the number of iterations, and the result reveals amazing images that reflect the hidden beauty of fractals.

As we are interested in creating an image, the complex plane is effectively the display device, and is divided into pixels defined by the required image resolution. Each pixel has an associated complex number $c = x + iy$, which is input to the algorithm and iterated. The result determines the pixel's colour.

The algorithm is extremely simple:

$$z_{n+1} = z_n^2 + c, \quad \text{where} \quad \{z, c\} \in \mathbb{C}, \quad \text{and} \quad z_0 = 0.$$

© Springer International Publishing AG, part of Springer Nature 2018
J. Vince, *Imaginary Mathematics for Computer Science*,
https://doi.org/10.1007/978-3-319-94637-5_13

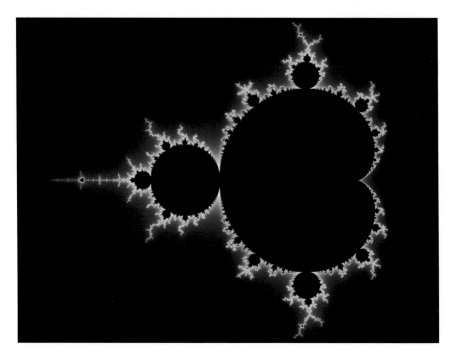

Fig. 13.1 The Mandelbrot set created by the program Ultra Fractal 3. Reproduced by kind permission of Dr. Wolfgang Beyer

By iterating this algorithm, subject to the two test criteria, the point c is classified as in $\mathbb{M}$ or not. The sequence of z_0, z_1, z_2 etc., is called the *orbit* of c. Some points only require a few iterations to show that they are unbounded, and the orbit moves off towards infinity. Some may oscillate between two small values and confirm that they are bounded. Others may take many iterations before the orbit reveals whether c is bounded or otherwise.

To start the algorithm, $z_0 = 0$, and c is any chosen complex pixel. For example, when $c = 0$, the value of z remains unmoved at 0, therefore $0 \in \mathbb{M}$. However, when $c = 1$, the value of z_n grows rapidly: $0, 1, 2, 5, 26, 677, \ldots, \infty$, is unbounded, and $1 \notin \mathbb{M}$. When $c = -1$, the value of z_n oscillates between -1 and 0, and is bounded by the set. This process is repeated for all values of c and the calculation repeated until $c \in \mathbb{M}$, $c \notin \mathbb{M}$, $|z_n| \geq 2$ or a specified number of iterations is exceeded.

Table 13.1 shows part of the orbit of z_n for $c = i$: $z_n = i, (-1 + i), -i,$ $(-1 + i), -i$ and is clearly bounded. Similarly, Table 13.2 shows part of the orbit of z_n for $c = -i$: $z_n = -i, (-1 - i), i, (-1 - i), i$ and is also bounded. Finally, the point $c = -2$, produces $-2, 2, 2, 2, \ldots$ and is also inside $\mathbb{M}$.

Figure 13.2 shows how $\mathbb{M}$ evolves from 10 to 80 iterations, and Fig. 13.3 shows the set after zooming in to the black circle on top of the main cardioid. There were 5000 iterations, which creates the fine detail.

Table 13.1 Mandelbrot algorithm for $c = i$

n	z_n	z_{n+1}
0	0	i
1	i	$-1+i$
2	$-1+i$	$-i$
3	$-i$	$-1+i$
4	$-1+i$	$-i$

Table 13.2 Mandelbrot algorithm for $c = -i$

n	z_n	z_{n+1}
0	0	$-i$
1	$-i$	$-1-i$
2	$-1+i$	i
3	i	$-1-i$
4	$-1-i$	i

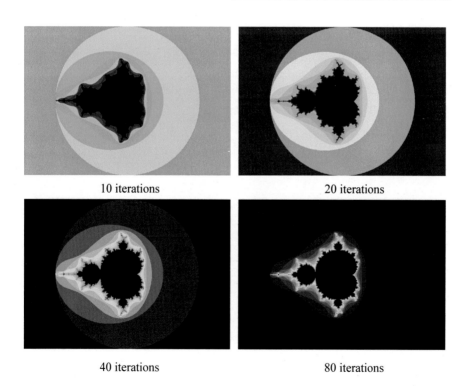

10 iterations   20 iterations

40 iterations   80 iterations

Fig. 13.2 The Mandelbrot set with increasing number of iterations. Reproduced by kind permission of Dr. Dominic Ford

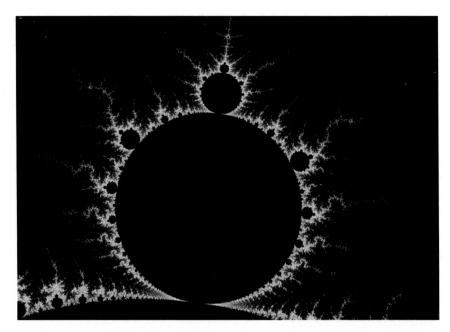

Fig. 13.3 Zooming in with 5000 iterations. Reproduced by kind permission of Dr. Dominic Ford

Fig. 13.4 Part of the Mandelbrot set. Reproduced by kind permission of Dr. Wolfgang Beyer

My thanks to Dr. Dominic Ford (www.sciencedemos.org.uk). and Dr. Wolfgang Beyer for permission to use their images. Last, but not least, I include an image by Dr. Beyer showing the endpoint of the "seahorse tail" downloaded from www.wikipedia.org (Fig. 13.4).

Chapter 14
Conclusion

14.1 Imaginary Mathematics for Computer Science

Studying computer science prepares you for a wide range of careers. You could start running a website design company, offering consultancy in cyber security or working as a technical director for a special-effects company – to name but three. Consequently, it is impossible to address every aspect of computer science relevant to these career opportunities. But what is possible, is to design a teaching programme containing the essential foundations that give breadth to new knowledge, and can be extended if necessary with higher education.

While writing this book, I tried to keep this an introductory text for complex numbers, with topics I hope are relevant to computer science. This has made certain subjects, such as quaternions, octonions, matrix algebra and geometric algebra essential subjects. Complementing these, I included circuit analysis, geometry, combining waves and 3D rotations, that show real applications for complex numbers. The two chapters on the Riemann Hypothesis and the Mandelbrot set, reveal how mathematicians have employed complex numbers to resolve challenging problems in prime numbers and fractals, both of which, are relevant to computer science.

If you are still interested in learning more about complex numbers, there is the subject of complex analysis, hypercomplex numbers in n dimensions, geometric algebra for physicists, etc. Finally, I include some famous complex equations that have created the new subject of quantum field theory. They may tempt you to delve deeper into this exciting subject!

14.1.1 Pauli Matrices

These matrices are named after the Austrian-born Swiss and American theoretical physicist Wolfgang Pauli (1900–1958), who used them in quantum mechanics in relation to the spin of atomic particles in an electromagnetic field.

© Springer International Publishing AG, part of Springer Nature 2018
J. Vince, *Imaginary Mathematics for Computer Science*,
https://doi.org/10.1007/978-3-319-94637-5_14

$$\sigma_1 = \begin{bmatrix} 0 & 1 \\ 1 & 0 \end{bmatrix}, \quad \sigma_2 = \begin{bmatrix} 0 & -i \\ i & 0 \end{bmatrix}, \quad \sigma_3 = \begin{bmatrix} 1 & 0 \\ 0 & -1 \end{bmatrix}$$

These matrices are Hermitian and unitary and

$$\sigma_1^2 \sigma_2^2 \sigma_3^2 = -i\sigma_1\sigma_2\sigma_3 = I$$

$$\sigma_1^2 = \begin{bmatrix} 0 & 1 \\ 1 & 0 \end{bmatrix} \begin{bmatrix} 0 & 1 \\ 1 & 0 \end{bmatrix} \qquad\qquad = \begin{bmatrix} 1 & 0 \\ 0 & 1 \end{bmatrix}$$

$$\sigma_2^2 = \begin{bmatrix} 0 & -i \\ i & 0 \end{bmatrix} \begin{bmatrix} 0 & -i \\ i & 0 \end{bmatrix} \qquad\qquad = \begin{bmatrix} 1 & 0 \\ 0 & 1 \end{bmatrix}$$

$$\sigma_3^2 = \begin{bmatrix} 1 & 0 \\ 0 & -1 \end{bmatrix} \begin{bmatrix} 1 & 0 \\ 0 & -1 \end{bmatrix} \qquad\qquad = \begin{bmatrix} 1 & 0 \\ 0 & 1 \end{bmatrix}$$

$$-i\sigma_1\sigma_2\sigma_3 = -i \begin{bmatrix} 0 & 1 \\ 1 & 0 \end{bmatrix} \begin{bmatrix} 0 & -i \\ i & 0 \end{bmatrix} \begin{bmatrix} 1 & 0 \\ 0 & -1 \end{bmatrix}$$

$$= -i \begin{bmatrix} i & 0 \\ 0 & -i \end{bmatrix} \begin{bmatrix} 1 & 0 \\ 0 & -1 \end{bmatrix} \qquad\qquad = \begin{bmatrix} 1 & 0 \\ 0 & 1 \end{bmatrix}.$$

14.1.2   Dirac Matrices

These matrices are named after the English theoretical physicist Paul Dirac (1902–1984):

$$\gamma^0 = \begin{bmatrix} 1 & 0 & 0 & 0 \\ 0 & 1 & 0 & 0 \\ 0 & 0 & -1 & 0 \\ 0 & 0 & 0 & -1 \end{bmatrix}, \quad \gamma^1 = \begin{bmatrix} 0 & 0 & 0 & 1 \\ 0 & 0 & 1 & 0 \\ 0 & -1 & 0 & 0 \\ -1 & 0 & 0 & 0 \end{bmatrix},$$

$$\gamma^2 = \begin{bmatrix} 0 & 0 & 0 & -i \\ 0 & 0 & i & 0 \\ 0 & i & 0 & 0 \\ -i & 0 & 0 & 0 \end{bmatrix}, \quad \gamma^3 = \begin{bmatrix} 0 & 0 & 1 & 0 \\ 0 & 0 & 0 & -1 \\ -1 & 0 & 0 & 0 \\ 0 & 1 & 0 & 0 \end{bmatrix}.$$

$$\gamma^5 = i\gamma^0\gamma^1\gamma^2\gamma^3 = \begin{pmatrix} 0 & 0 & 1 & 0 \\ 0 & 0 & 0 & 1 \\ 1 & 0 & 0 & 0 \\ 0 & 1 & 0 & 0 \end{pmatrix}.$$

γ^5 is Hermitian, and because

$$(\gamma^5)^2 = \begin{pmatrix} 0 & 0 & 1 & 0 \\ 0 & 0 & 0 & 1 \\ 1 & 0 & 0 & 0 \\ 0 & 1 & 0 & 0 \end{pmatrix} \begin{pmatrix} 0 & 0 & 1 & 0 \\ 0 & 0 & 0 & 1 \\ 1 & 0 & 0 & 0 \\ 0 & 1 & 0 & 0 \end{pmatrix} = \begin{pmatrix} 1 & 0 & 0 & 0 \\ 0 & 1 & 0 & 0 \\ 0 & 0 & 1 & 0 \\ 0 & 0 & 0 & 1 \end{pmatrix} = I_4$$

its eigenvalues are ± 1. The matrices also anti-commute.

14.1.3 The Dirac Equation

The above gamma matrices were used by Dirac in this equation, which predicts the existence of antimatter:

$$\left[\gamma^\mu \left(i \frac{\partial}{\partial x^\mu} - e A_\mu(x) \right) + m \right] \psi(x) = 0, \quad \text{where} \quad 0 \leq \mu \leq 3.$$

γ^μ references one of the four gamma matrices, γ^0, γ^1, γ^2, γ^3,
$\psi(x)$ is a wave function with four components $\psi_{e\uparrow}(x)$, $\psi_{e\downarrow}(x)$, $\psi_{p\uparrow}(x)$, $\psi_{p\downarrow}(x)$
$e\uparrow$ and $e\downarrow$ stands for an electron with spin up and spin down, respectively,
$p\uparrow$ and $p\downarrow$ stands for a positron with spin up and spin down, respectively,
$\partial/\partial x^\mu$ measures the rate at which the wave function is changing in time and space,
$A_\mu(x)$ represents the electromagnetic field potentials,
$-e$ is the electron's electric charge,
m is the electron's mass.

In Dirac's original equation:

$$\left(\beta m c^2 + c \left(\sum_{n=1}^{3} \alpha_n p_n \right) \right) \psi(x, t) = i\hbar \frac{\partial \psi(x, t)}{\partial t}$$

Prof. Hestenes identified the term $i\hbar$ as a bivector, which describes an oriented plane in geometric algebra.

14.1.4 The Schrödinger Wave Equation

The Austrian theoretical physicist Erwin Schrödinger (1887–1961), is famous for many things, but most notably his wave equation, and his imaginary cat experiment. Here is his equation:

$$i\hbar \frac{\partial}{\partial t}\psi(r, t) = -\frac{\hbar^2}{2m}\nabla^2\psi(r, t) + V(r, t)\psi(r, t)$$

$\hbar = h/2\pi$ is called the Dirac constant, where h is Planck's constant,
$\psi(r, t)$ is the wave function in space and time,
$\frac{\partial}{\partial t}\psi(r, t)$ is the rate of change of the wave function,
m is the mass of a particle,
$\nabla^2\psi(r, t)$ is the Laplacian of the wave function, where

$$\nabla = \frac{\partial}{\partial x}\mathbf{i} + \frac{\partial}{\partial y}\mathbf{j} + \frac{\partial}{\partial z}\mathbf{k}$$

and

$$\nabla^2\psi = \frac{\partial^2\psi}{\partial x^2} + \frac{\partial^2\psi}{\partial y^2} + \frac{\partial^2\psi}{\partial z^2}$$

$V(r, t)$ is the potential energy of the system.

Quantum field theory computes the probability of finding a particle in a place and time, which is derived from the square of the Euclidean norm of the wave function.

14.2   The Imaginary Unit

All the above equations contain the *imaginary unit* $i = \sqrt{-1}$, which initially resolved solutions to quadratic equations, and today is being used in quantum field equations.

Index

© Springer International Publishing AG, part of Springer Nature 2018
J. Vince, *Imaginary Mathematics for Computer Science*,
https://doi.org/10.1007/978-3-319-94637-5